# City Hospitals

# City Hospitals,
## The Undercare of
## the Underprivileged

HARRY F. DOWLING

Harvard University Press

Cambridge, Massachusetts
and London, England 1982

Library of Congress Cataloging in Publication Data

Dowling, Harry Filmore, 1904–
  City hospitals.

  Includes bibliographical references and index.
  1. Hospitals, Public—United States—History.
2. Hospital care—United States—History. 3. Urban
health—United States—History. I. Title. [DNLM:
1. Hospitals, Public—History—United States. 2. Quality
of health care—United States. WX 27 AA1 D7c]
  RA981.A3D68      362.1'1      82-1069
  ISBN 0-674-13197-5            AACR2

# Acknowledgments

During the period when this book was being written, I was a visiting scholar at the National Library of Medicine, and the facilities of that library were placed at my disposal by Martin Cummings and his staff. John Blake and James H. Cassedy of the Division of the History of Medicine offered valuable advice and criticisms throughout the undertaking and read the manuscript. Others who read the manuscript and gave helpful advice were: Lloyd G. Stevenson and Edyth Schoenrich of Johns Hopkins University, Jon M. Kingsdale of Harvard University, William G. Rothstein of the University of Maryland, and Peter Olch and Manfred Waserman of the National Library of Medicine. Readers of the manuscript who represented the physicians' viewpoint included Edmund G. Beacham of Baltimore City Hospitals; John J. Byrne and Maxwell Finland of Boston City Hospital; Rolf M. Gunnar, Hugh H. Hussey, and Sol Katz, formerly of Cook County and District of Columbia General Hospitals; and Alice B. Tobler, formerly of the Health Department of the State of Maryland. My son William Laine Dowling, of the University of Washington, read the manuscript from the standpoint of the hospital administrator.

Others who read portions of the manuscript or gave information or advice included: C. Holmes Boyd, Charles C. J. Carpenter, William B. Castle, Francis Chinard, W. Montague Cobb, Caroline Cochran, Nicholas J. Cotsonas, Jr., J. Keith Cromer, Susan Cromer, Charles Davidson, Burton d'Lugoff, H. Filmore Dowling, Jr., Franklin Epstein, Harry A. Feldman, Joseph Gonnella, Anne G. Hargreaves, A. M. Harvey, Harold Jeghers, Esther Lazarus, John M. Leedom, Joyce S. Lepper, Mark H. Lepper, William R. McCabe, Max M. Montgomery, Mary Mullane, Jack Myers, David Rogers, Max Schmidt, Patrick Storey, Marvin Turck, Thomas B. Turner, Robert Williams, Theodore Woodward, Philip Zieve, and Hyman J. Zimmerman.

I am indebted to librarians and archivists at a number of institutions, especially Lawrence Brenner of Boston City Hospital, Dorothy Hanks of the National Library of Medicine, Kathy Jacob, archivist of the United States Senate, Adlea Jones of the Baltimore City Hospitals, Nancy McCall of the Alan Mason Chesney Archives of Johns Hopkins University, Terence Norwood of Cook County Hospital, and Richard Wolfe of the Francis A. Countway Library of Medicine.

Finally, I appreciate the ever-present help of Mary Elizabeth Rose, who uncovered information from sources near and far—in time and in location—assembled data, and edited the manuscript.

This publication was supported in part by NIH Grant LM00008 from the National Library of Medicine and in part by a grant from the Robert Wood Johnson Foundation.

# Contents

# City Hospitals

# Introduction

There ghastly Diseases dwell, and joyless
Old Age, and Fear, and Hunger...and
squalid Want...

Vergil, *Aeneid*

NOBODY WANTS TO be sick, or if sick to go to a hospital if he can help it; nobody wants to be poor, or if poor to be a public charge if he can avoid it. Yet many are sick and poor and dependent on others for help. In the United States a special set of institutions evolved to care for the sick poor—the general hospitals owned and operated by local governments. For over a century these hospitals have played a major role in the care of the sick, the education of health professionals, and the furtherance of medical research in the United States, but since most people try to forget sickness and shut their eyes to poverty, society has largely ignored these insitutions and knows little about them. The story deserves to be told how a few dedicated people, operating with too few resources, struggled against great odds to develop a workable system for the institutional care of the underprivileged sick. Furthermore, since changing circumstances now demand that the system be modified, the American people need to know the story of the past when they are being called upon to decide how to provide health care for the sick poor and others in the years to come. For we are told in the parable that the Samaritan first had to stop and learn before he could take the action needed to become a good neighbor.

Hospitals caring for patients with the general run of short-term illnesses and owned by local or state governments have been designated public-general hospitals. Their primary purpose is to render medical care to those who need it but cannot pay for it either because they are without the means to do so or have been rendered unable to pay by the illness itself. Public-general hospitals also admit other patients: the traveler, whether a sailor from a far country or a traffic victim from the next county; wards of the government who become ill—orphans, prisoners, aged and otherwise helpless persons; and in recent years, for various reasons and to varying degrees, persons who can pay, or whose insurance will pay, for their care.[1]

This book focuses on publicly owned general hospitals in large metropolitan areas of the United States. They may be owned by a city, a county, a regional authority, or, occasionally, by a state when its hospital is the sole public institution serving as a general hospital for a large metropolitan area and thus functioning as a hospital for the city. (An example of the last is Charity Hospital in New Orleans.) In the nation's one hundred largest cities there are ninety public-general hospitals, excluding university-owned ones, which deliver 13.2 percent of all inpatient services and 28.9 percent of all hospital outpatient services in these cities.[2] I have chosen to call them by the short, familiar, and historic term *city hospitals,* because they function as the main public hospitals for the poor of the cities even though they may be owned by some other governmental authority.

Locally owned public hospitals in smaller cities and rural communities tend to differ from city hospitals not only in size but, more important, in the kinds of patients they care for. These hospitals admit a substantial number of private patients, and thus in many respects they resemble private hospitals. The latter are hospitals that are not owned by any unit of government. They include voluntary hospitals, those owned by nonprofit voluntary organizations, and proprietary hospitals, operated for profit by corporations, partnerships, or individuals.[3]

The history of city hospitals in the United States can be divided into four periods. In the poorhouse period, which lasted until approximately 1860 in the larger eastern cities and somewhat later in many younger cities, the hospital was considered an unimportant part of the almshouse and subordinate to it; the general welfare functions of the almshouse tended to predominate over medical care. In the practitioner period providing medical and associated nursing care became the dominant purpose of the city hospitals; medical care was rendered by doctors who took time off from their practices for this purpose, with minimal assistance from hospital-based doctors, and nursing care evolved with the opening of hospital schools of nursing. The start of the academic period was marked by the appearance in 1910 of Abraham Flexner's report on medical education in the United States and Canada.[4] This period saw the installation of full-time medical staffs controlled by medical schools, organized instruction of medical students as part of the medical care team, the flowering of hospital-based nursing schools, and the rise of collegiate schools of nursing. In the 1960s the emphasis began to shift from developing city hospitals as academic medical centers to organizing them as leaders of health care for their communities. I have set the date for the beginning of the community period at 1965, when the far-reaching amendments to the Social Security Act began to operate. The chapter dealing with this period is primarily an epilogue, for the transition is still very much in progress. In each period it is important to understand those factors within and without city hospitals that relate to governance and administration; medical care, education, and research; and other elements affecting patient care, especially nursing services and nursing education.

As examples of city hospitals during the poorhouse period I have used those in the older eastern cities, particularly the pioneer Philadelphia (later Philadelphia General) Hospital and Bellevue Hospital in New York. For subsequent periods the prototypes used are the city hospitals in Baltimore, Boston, Chicago, and Washington, with each of which I have been associated at some time. Personal observation and experience have brought knowledge and insights that could not have been obtained in any other way. I have also had access to primary as well as secondary sources and have been able to question present and former administrators, staff members, and other health professionals. When one is dealing with events of recent decades and with personal experiences, it is difficult to be wholly objective. I have attempted to avoid partiality by seeking information from various sources and by trying to draw conclusions that are fair to all.

Data on other hospitals were gathered from periodicals, the few histories available, and conversations with colleagues who had been associated with them. Generalizations made from the detailed information obtained on the prototype hospitals have rarely required modification as a result of data obtained from other hospitals; rather, information from the latter has almost always served to verify conclusions reached from information obtained on the prototype hospitals.

I began this project with the attitude, shared by most doctors who have worked in city hospitals, that if only the public were willing to provide enough money, and if only the politicians and their incompetent henchmen would get out of the way, doctors would see to it that patients receive good care. I soon saw that the problems are more complicated, that both builders and wreckers are to be found in every group, and, strange to relate, that doctors themselves are sometimes the problem. It became necessary to study outside actions and events that affect the fate of city hospitals, such as activities of city governments, medical schools, and the federal government; changes in other professions, especially the nursing, social service, and hospital administration professions; and finally the attitudes of the public.

In recent years city hospitals in the United States have been going through a trying period, perhaps the most difficult in their long history. They may not even survive the next few years; if they do they will surely be greatly changed. Because both their troubles and their triumphs have deep roots, an interpretation of their present status in the light of their past history should help the public and those directly concerned with health care to make decisions about their future role. Thus, this book is written for both of these groups.

# Part I · *The Almshouse Period*

# 1 · *Over the Hill from the Poorhouse*

Special care is first taken of the sick...in
public hospitals. They have four at the city
limits,...very well furnished and equipped
with everything conducive to health. Besides, such tender and careful treatment
and such constant attendance of expert
physicians are provided that, though no one
is sent to them against his will, there is
hardly anybody in the whole city who,
when suffering from illness, does not prefer
to be nursed there rather than at home.
Sir Thomas More, Utopia

In fine, it is a Hospital for Curables and
Incurables of all ages and sexes, and in
every Disease and Malady, even to Lunacy
and Idiotism...; and very generally an Institution for clothing the Naked, feeding
the hungry, healing the sick and administering Comfort and Relief to the distressed of
every kind.
Description of Philadelphia almshouse,<br>1775

Not infrequently those who are the most
deserving objects of charity cannot be in-
duced to enter it. To some of them, death
appears less terrible than a residence in the
almshouse.
Comment on Boston almshouse, 1810

TODAY'S CITY- AND county-owned general hospi-
tals in the United States can be traced directly to sixteenth-century
England. The closing of religious institutions by Henry VIII in the
1530s threw the burden of supporting the poor and sick on the public,
and between 1552 and 1601 a series of acts of Parliament placed this
responsibility upon the local communities. Local officials found
themselves responsible for the aged, chronically ill, and permanently
disabled; homeless women before and during childbirth; foundlings
and orphans; the insane, the feeble-minded, and alcoholics; and
persons with medical and surgical illnesses who today would be placed
in a general hospital. To prevent what they saw as the spread of
pauperism, the towns and cities refused aid to people in their homes.
Instead, they packed all these groups into almshouse-workhouse

complexes, along with vagrants and petty criminals. Here they could force the able-bodied to earn their keep and house cheaply those unable to work. According to Charles Dickens, everyone had the option of "being starved by a gradual process in the house, or by a quick one out of it."[1]

Unlike the care of the sick in Utopia, conditions in many of these institutions were deplorable. A report in 1834 stated: "The sick were housed with prostitutes, ex-gaol birds, tramps, and other characters of the worst description, and their sufferings were not infrequently aggravated by the incessant ravings of some neglected lunatic." Usually the sick in the almshouse were visited, under a contractual arrangement, by a doctor who practiced in the nearby community. They were nursed by inmates—old women still able to hobble about, prostitutes picked up for soliciting or vagrancy, or others convicted of petty offenses.[2] Under the circumstances medical attention was sometimes casual and inept, while nursing was often slovenly, callous, even cruel. In a typical workhouse in 1766 the infants

> were put into the hands of indigent, filthy, or decrepit women, three or four to one woman, and sometimes sleeping with them. The allowance of these women being scanty, they are tempted to take part of the bread and milk intended for the poor infants. The child cries for food, and the nurse beats it because it cries. Thus with blows, starving and putrid air, with the additions of lice, itch, and filthiness, he soon receives his quietus.[3]

In some workhouses infant mortality reached 82 percent. From 1750 to 1755, in fourteen of the largest parishes, among 1,265 children who remained in almshouses more than a few days, only 168 were living at the end of the period.[4]

Improvement was slow. A century later an inspector of the Poor Law Board reported of one institution that there were not enough beds; dishes were dirty; necessary utensils and supplies were inadequate and food badly cooked and cold; doctors were seldom called and then were poorly paid; and "the patients frequently...were nursed by paupers...whose love of drink often drove them to rob the sick...and whose treatment of the poor was, generally speaking, characterized neither by judgement nor by gentleness." This report resulted in the Metropolitan Poor Act of 1867, which is considered the beginning of modern poor relief in England.[5]

As early as 1123 another kind of institution had arisen in England: the voluntary hospital, supported by private charity and run by an independent board of prominent citizens. For some time after the first of these, St. Bartholomew's, was founded, they were few in number and confined to London, but between 1710 and 1797 several more were established in London and twenty-seven in the provinces. They differed from almshouses and the hospitals that evolved within them in ways other than governance and source of funds. They were more selective, preferring to admit acutely ill patients who might be expected to recover, those recommended by one of the subscribers, or those who could pay something toward their care. Leading doctors were attracted to the staffs of the voluntary hospitals through social acquaintance with lay members of governing boards and because appointments carried prestige and offered opportunities to teach medical students. Careful selection of patients, a superior attending staff, the presence of medical students, and close attention by the lay board all tended to raise morale and improve the quality of care given to patients. Thus, from the beginning voluntary hospitals generally gave far better care than almshouse hospitals. They—not the public hospitals—resembled Thomas More's ideal institution for the care of the sick.[6]

In the British colonies of North America hospitals developed slowly because the population was predominantly rural. In 1700 Boston had 6,700 inhabitants, New York about 4,500, and Philadelphia 4,400. On the eve of the Revolution only five cities contained more than 8,000 inhabitants each, with fewer than 100,000 in all five. By the early nineteenth century towns and cities were growing rapidly. In 1790 the number of cities with populations over 8,000 was still only five—New York, Philadelphia, Baltimore, Boston, and Charleston (see Table 2 in Chapter 2 for populations of some of these cities)—containing altogether sightly over 130,000 inhabitants; by 1820 there were thirteen with a total population of nearly half a million. The first charters to be granted to a borough or city were given to New York and Albany in 1686 and to Philadelphia in 1691. New York had a borough government after the Dutch model as early as 1653, and this became an English municipal corporation under English rule in 1686.[7]

As cities grew, poverty intensified, especially after wars and during slumps in business cycles. Crowding, filth, hunger, and lack of medical care among the poor increased and compounded their illnesses, and epidemics inevitably hit the poorest sections hardest. Added to the native poor were the waves of immigrants, many of whom arrived "half starved, half sick, and often barely alive." The few funds they had were soon gone, often taken by thieves or confidence men. When they were able to find work, their scanty wages forced them to live on a subsistence diet or less. Thus, immigrants were especially susceptible to diseases of dirt and deprivation—typhoid, dysentery, typhus, scurvy, rickets, and tuberculosis.[8]

The British colonies in America adopted the English dual system of hospitals. Groups of public-spirited citizens founded voluntary hospitals, such as the Pennsylvania Hospital, which opened in Philadelphia in 1751 or 1752, the New York Hospital in 1791, and the Massachusetts General Hospital in Boston in 1821. Other hospitals were established by religious groups and by individuals. But these cared for only a fraction of the poor who needed hospitalization; most of the indigent sick who were admitted to any institution entered the publicly owned almshouses. These usually functioned also as work-houses and often as jails, insane asylums, and homes for foundlings and orphans, once again imitating the English system. For instance, a special committee on pauper laws, reporting to the Massachusetts General Court in 1811, emphasized that the evil of pauperism could be most effectively diminished "by making Alms Houses, Houses of Industry, and not abodes of idleness, and denying for the most part all supply from public provision, except on condition of admission into the public institution."[9]

Almshouses had been established as early as 1612 at Henricopolis in Virginia, in 1658 in New Amsterdam, and in 1665 in Boston. These almshouses did not become hospitals, even though one of their functions was to care for the sick. In Boston, for instance, the almshouse was for a long time the only facility in which the physically and mentally ill could be placed. Yet its hospital function was not stressed: in 1814 one reason given for the need for Boston's first voluntary hospital, the Massachusetts General, was that the almshouse contained accommodations for only eight patients.[10] Instead of being part of the almshouse, Boston's first public-general hospital, when it was established in 1864, was an independent institution.

The earliest public almshouse that evolved into a hospital was opened in Philadelphia in 1731 or 1732 (see Table 1). Doctors apparently treated patients there from the beginning—not only inmates who became ill while in the almshouse but also patients admitted directly for therapy. Records show that Dr. William Shippen was paid thirty pounds for medicines provided to almshouse patients in 1738 and 1739, a fact that clearly establishes medical attendance at that institution. Furthermore, the managers of the almshouse, addressing the Assembly of the Province of Pennsylvania in 1775, claimed it was "not only an Asylum for the poor, old and emaciated, as an Alms-House; but...really and fully an Hospital, in every sense of the word." In 1795 a count showed that 114 of the 301 inmates of the institution were ill with a variety of diseases. Thus, this almshouse functioned as a hospital from its founding or soon after; it eventually became the Philadelphia General Hospital.[11]

In 1834 the almshouse moved to 187 acres of farmland west of Schuylkill River in Blockley Township. From then on the name Blockley clung to almshouse and hospital alike—spoken fondly by doctors who trained or taught there but with varying degrees of fear

Table 1. Some early almshouse hospitals.

| | Philadelphia | New York | Baltimore | Washington, D.C. |
|---|---|---|---|---|
| Founding of almshouse | 1731 or 1732 | 1736 | 1776 | 1806 |
| Establishment of almshouse at present site | 1834 (Philadelphia Almshouse—Blockley) | 1816 (Bellevue Establishment) | 1886 (Bay View Asylum) | 1839 (Washington Asylum) |
| Nonhospital functions moved away | 1874–1926 | 1836–1848 | 1901–1964 | 1903–1916 |
| Designated as a hospital | 1835 (Philadelphia Hospital) 1902 (Philadelphia General Hospital) | 1825 (Bellevue Hospital) | 1925 (Baltimore City Hospitals) | 1914 (Gallinger Municipal Hospital) 1953 (District of Columbia General Hospital) |
| Hospital closed | 1976 | — | — | — |

or horror by former and prospective patients. Despite the abundance of land, the institution functioned in four massive brick and stone buildings placed around a quadrangle containing a few small frame structures, all surrounded by a high board fence. Although the various activities were carried on in different buildings, the institution operated as a single unit. Separation really began in 1835, when the hospital portion was designated the Philadelphia Hospital. Yet the care of psychiatric and almshouse patients was still a reponsibility of the hospital department until 1859.[12]

It was a long time before the city authorities decided that inmates who did not need a general hospital might be better off in a more rural setting. A House of Correction beyond the city was opened in 1874. In the 1880s pressure developed to move the almshouse and psychiatric hospital outside the city, but it was much cheaper to transfer them to remodeled buildings on Blockley grounds, and this was done beginning in 1906. In 1914 some of these patients were removed from the hospital campus, but not until 1926 were the last ones taken away.[13]

In New York City the House of Correction, Workhouse and Poorhouse that opened in 1736 eventually became Bellevue Hospital (see Table 1). Like the Philadelphia almshouse, it cared for the sick from the start, having a competent attending physician, John Van Buren, who had been a student of the famous medical teacher Hermann Boerhaave. In 1739 the hospital function was further extended when a separate building was erected for patients with contagious diseases. Yet a prison atmosphere was present also in the iron cages for psychiatric patients and whipping posts for slaves sent by their owners to be punished. The official in charge was called the Keeper of the House of Correction and the Master of the Workhouse and Poorhouse.[14]

When the Bellevue complex was established at its present site in 1816, it included an almshouse, workhouse, penitentiary, and two separate buildings for the male and female sick. The growth of the hospital portion went largely unnoticed. In 1825 a committee of doctors reminded the Common Council that "whereas they imagined they were conducting but an almshouse and a prison," they "in reality were administering a large hospital." This report convinced the authorities that they should concentrate on the care of the sick. In that year they designated the hospital portion Bellevue Hospital, and a few years later they began to move the other divisions away: the male prisoners in 1836, next the female prisoners, and finally in 1848 the almshouse.[15]

Almshouse infirmaries in other large cities grew into city hospitals. The nations's capital in the District of Columbia, in contrast to most eastern cities, is a young city; its site consisted mainly of farmland until the federal government moved there in 1800.[16] In 1806 the first almshouse, the Washington Infirmary, was erected and at the time was the only hospital to which the indigent sick could go (see Table 1). In 1839 a larger almshouse and workhouse, the Washington Asylum, was built two miles southeast of the Capitol on the site of the present District of Columbia General Hospital.[17] At first the jail and poorhouse shared the same building, separated by a thin board partition. Later they were placed in separate but neighboring buildings. In 1841 the first of a series of small frame buildings was erected nearby to house smallpox patients. Other frame buildings were added, particularly during the Civil War, when ill and wounded soldiers were added to the institution's population. These many structures enabled the superintendent, who controlled jail, poorhouse, and hospital, to keep the various groups separate, and thus the complex continued into the twentieth century. As in Philadelphia and New York, a general hospital developed out of necessity, and like theirs it was left behind as the poorhouse-workhouse complex was moved away between 1903 and 1916. The name was changed from Washington Asylum Hospital to Gallinger Municipal Hospital in 1914.[18]

Baltimore's first almshouse, which operated from 1776 to 1822 on a site just northwest of the present business district, "served all the needs of the indigent of Baltimore. It was an acute, chronic, and lying-in hospital, a workhouse, a shelter for vagrants, criminals awaiting trial, the insane, foundlings, and the old and infirm" (see Table 1).[19] So was its successor at Calverton in the western part of the city (1822–1865) and the present institution in east Baltimore, which opened in 1866 under the name of Bay View Asylum. At the end of the nineteenth century one administration was still responsible for the care of a heterogeneous group of acute and chronically ill, alcoholics, psychotics, and petty criminals. Some time before 1831 patients with medical and surgical diseases were moved to their own wards, and in 1897 senile patients were assigned to a separate annex. Not until 1911 were patients with acute illnesses moved to a different building equipped as a modern hospital and patients with tuberculosis placed in their own buildings, as psychiatric patients already had been. In 1925 the shift toward a general hospital was formally proclaimed when the institution's name was changed to Baltimore City Hospitals. Still, the institution remained a catchall until 1935, when magistrates were prohibited from committing suspected alcoholics and others to the city hospital.[20]

Thus, by the beginning of the nineteenth century many, probably most, of the institutionalized sick poor in several large cities in the East and elsewhere were cared for in almshouse infirmaries. Because the majority of the inmates were not in the hospital department, governing boards and administrators tended to emphasize the discipline and regimentation necessary to control the vagrants and petty criminals while providing the bare necessities of life to the aged and disabled and extracting as much work as possible from all inmates. They approached the care of sick people as an additional burden to be managed somehow. The stigma of the pauper and the criminal branded the entire institution, including the hospital. When Drs. James Jackson and John C. Warren were trying to arouse interest in building the Massachusetts General Hospital, they pointed out the defects of the almshouse as a hospital: "The institution, then, is made to comprehend what is more properly meant by an almshouse, a . . . house of correction, and a hospital. Now, the economy and mode of government cannot possibly be adapted at once to all these various purposes."[21]

Transformation into hospitals was gradual. Facilities for surgery and for deliveries were installed and improved step by step. Wards for the acutely ill were enlarged and these patients separated as much as possible from those having different needs and requirements: prisoners, the aged and permanently disabled, and the insane. Patients with tuberculosis, with contagious diseases, and with mental illnesses were placed in separate wards and then in separate buildings. In this way the hospitals, having established their separate identities by the mid-nineteenth century, loosened their ties with almshouses by degrees. The pattern was similar for younger cities, such as Washington and many midwestern and western cities. There, although the first almshouses were established later, the sequence was telescoped into a shorter period.

We know much more about when and where almshouse infirmaries were built than we do about the care given to the patients. We would like to believe the accounts of lay visitors to almshouses and their hospital wards, which were highly favorable, even glowing. Such was the Reverend Manasseh Cutler's description of Benjamin Rush's rounds at the Philadelphia almshouse in 1787:

> We returned to the upper hall, where several physicians and all the young students in physics [*sic*] in the city were waiting. . . We entered the upper chamber of the sick, . . . a spacious room finely ventilated with numerous large windows on both sides. . . The room was exceedingly clean and nice, the beds and bedding appeared to be of good quality, and the most profound silence and order were preserved upon the doctor's entering the room. . . [S]uch is the elegance of these buildings, the care and attention to the sick, the spacious and clean apartments and the perfect order in everything, that it seemed more like a palace than a hospital, and one would almost be tempted to be sick, if they could be so well provided for.[22]

Let us look more closely at this idyllic scene. The respect accorded Benjamin Rush is not surprising; he was the leading medical practitioner of Philadelphia as well as a signer of the Declaration of Independence.[23] A visitor might have been similarly impressed by rounds at the almshouse of other eminent doctors, including William Shippen, probably the first to be chosen as an attending physician; Thomas Bond and Cadwalader Evans, who taught the first medical classes (in obstetrics) at the almshouse; William W. Gerhard, who, while a member of the attending staff in 1835 and 1836, delineated the differences between typhus and typhoid fevers;[24] and, toward the end of the almshouse period, Samuel D. Gross and D. Hayes Agnew, who did much to establish the reputation of American surgery abroad.[25]

The New York almshouse also had some outstanding doctors, including Valentine Seaman, who gave the first lectures there in obstetrics in 1799; David Hosack, who in 1808 was the first in America to tie the femoral artery for aneurysm; and John Van Buren, who in 1849 instituted the first public surgical clinics. In Baltimore the almshouse was a magnet that drew younger doctors who had received good training but as yet had found little opportunity to use it. For instance, Andrew Wiesenthal, the son of Baltimore's leading physician, returned in 1789 from three years' training at St. Bartholomew's Hospital in London and immediately became an attending physician at the almshouse. In the early nineteenth century a succession of able doctors attended there. Three were founders of the Washington Medical College, and two had received training in prestigious schools abroad.[26]

Not all the attending doctors, however, were of the caliber of Benjamin Rush, David Hosack, Andrew Wiesenthal, or the famous patriot Joseph Warren, who attended at the Boston almshouse from 1769 to 1773. There were not that many well-trained doctors.[27] Historians do not usually list the names or deeds of the poorer ones, but we can deduce that attending staff positions were not always eagerly sought when we find that between 1788 and 1805 the average term of service at the Philadelphia almshouse was less than four years. Apparently in those days, as in city hospitals later, attending at the almshouse hospital was merely a steppingstone to more desirable positions. For all the well-trained and dedicated doctors there must have been dozens who had a shorter and less effective education. Some entered practice after spending time with an incompetent or irresponsible preceptor who took their money and taught them little, others after listening to a few lectures but without seeing a patient. Almshouses must have had to hire some doctors of this stamp merely to get the day's work done.[28] Furthermore, when the medical affairs of a hospital were under political control—and this happened all too often—doctors with more ambition than ability often received appointments. In fact, in the public's mind the difference between what laymen and doctors knew was so slight that in the early days of the Philadelphia almshouse some laymen volunteered to serve as doctors and were allowed to do so. Unfortunately, we have no record of how their patients fared.[29]

By the end of the eighteenth century the almshouse hospitals were beginning to appoint resident doctors, who performed therapeutic procedures ordered by the attending doctors and substituted for them in their absence. A resident apothecary was appointed at the Philadelphia almshouse in 1788 and a second one the following year. The Baltimore almshouse had a resident physician by 1818, and Bellevue had one with two assistants by 1825. They obtained histories from and examined patients, cupped and bled them and dressed their wounds, prepared the surgical instruments, and compounded the medicines.[30]

These resident physicians varied in ability and dedication, but at least their mistakes could be corrected by more experienced attending physicians and surgeons, and a system evolved that functioned reasonably well. Trouble came when politicians seized control of almshouse hospitals. This occurred at Bellevue, for instance, during a fifteen-year period beginning in 1832. To tighten their hold on the hospital, the party in power eliminated the attending staff and appointed their own man to be in charge of medical care. This chief resident physician was usually inexperienced and overworked, since his only assistants were medical students. Finally, in 1847 simultaneous epidemics of typhus and typhoid fevers occurred. During the ten worst weeks 1,226 patients with fevers were admitted. This additional burden revealed the ineptitude of the existing administration. City authorities were forced once more to appoint an attending staff, and some of the most prominent doctors of the city agreed to serve. Other reforms were instituted including examinations for house officers. Medical care improved so much that the mortality rate in the hospital, which had averaged 20 percent in the twenty years before 1847, fell to 10 percent by 1850.[31]

At the Philadelphia Hospital also a tug-of-war developed between the medical staff and the politicians over the control of medical care. In 1845 the house officers several times protested the quality of the food served them. When a cockroach crossed their dining table, they demanded to be seated with the matron, since she was served better food, presumably free of vermin. This privilege was refused, and they resigned, the medical staff siding with them. The governing board responded by firing the attending staff and appointing a chief resident to oversee the residents and manage patient care, and the teaching of medical students was stopped. A few years later the attending staff was again placed in charge of medical care and teaching, but in 1857 the board abolished teaching a second time, and the attending staff resigned. A truce was arranged the next year, but not until 1859, when an appointed board of guardians replaced the elected, politically motivated board, was authority over professional matters finally delegated to the attending staff.[32]

Even the best of doctors could do little to alter the course of disease during this period. Operations were infrequent. From May 1834 through April 1835, 2,571 patients were admitted to the hospital wards of the Baltimore almshouse. Only 104 operations were performed, most of which were minor, such as tapping for dropsy, circumcision, and the removal of small, easily accessible tumors. Specific remedies, medicines that act directly on diseases or their underlying causes to cure or ameliorate them, were almost nonexistent.[33] Thus, the therapy needed by most patients was bed rest and nursing care, provided, of course, that the nurses were skillful, attentive, and considerate, the bedclothes clean, the diet proper, and the food well prepared. But were they?

The Reverend Mr. Cutler inferred that the nurses at the Philadelphia almshouse were efficient because he observed that the wards where rounds took place were neat and clean and because of the "care and attention to the sick."[34] We would like to believe that this scene was typical of what went on every day and in other wards of the almshouse, but any intern who has seen nurses scurrying around to get ready for the chief's rounds would doubt it. What happened at other times and in other wards?

Actually, poor nursing was to be found in nearly every hospital and particularly in almshouse hospitals, which used as attendants many inmates from the workhouse or jail and convalescing patients. These were often followers of Sarah Gamp, more familiar with gin bottles than medicine bottles and more given to filching money from patients' pockets than to changing their bedclothes. In 1809 an investigating committee at the Philadelphia almshouse found that attendants were selling coffee intended for patients and were habitually tipsy from drinking the liquor ordered for them. When drugs were prescribed, they gave the wrong ones or none at all and topped everything by stealing the clothing of patients who died.[35]

Epidemics of dysentery, typhus, and cholera threw an intolerable burden on an already overworked staff. Agnew related that during a cholera epidemic in 1832 the nurses at the Philadelphia almshouse "were seized with a kind of mad infatuation. They drank the stimulants provided for the sick, and in one ward . . . these Furies were seen lying drunk upon, or fighting over the dead victims of the disease." Propriety was not restored until nurses from the Sisters of Charity, a Roman Catholic order, were called in.[36]

Bellevue may have been at its lowest ebb in 1837, when a corrupt administration along with an epidemic of typhus reduced the institution to a shambles. The matron, the resident physician, with several of his students, and most or all of the nurses and helpers fled. Patients lay in their filthy blankets without sheets and pillowcases; no clothing could be found for them, no cornmeal for poultices, not even rags for dressings.[37] Nursing care remained generally poor at Bellevue throughout the almshouse period. In 1860, at the end of the period, the *New York Times* reported that a mother gave birth to a child during the night without receiving any attention, and that when the doctor saw her on his rounds the next day he found the baby dead and parts of its nose, cheeks, and left foot eaten off by rats.[38] Crowding was a problem that haunted doctors, nurses, and administrators in almshouse hospitals most of the time. In May 1847 (not the busiest time of the year) Bellevue contained 707 patients in beds or cots and 140 lying on the floor.[39] Such conditions were common and were worse in the winter months and during epidemics.

Poor food was another hazard for patients. At the Philadelphia almshouse in 1784 the food was found to be filled with maggots. The standard diet established in 1804 consisted of one pound of potatoes, one-quarter pound of meat, and three-quarters of a pound of bread per day, which would provide about three-quarters of the minimal number of calories now considered adequate. At times in the mid-nineteenth century the diet in the hospital department was so scanty that patients were constantly hungry, and scurvy flourished. And almost always the food was poorly cooked and sloppily served.[40]

Although inferior personal care in the almshouse hospitals was in part the fault of untrained, unkempt, and uninterested attendants, the main causes of poor and scanty food, shortages of bedclothing and supplies, and the general shabbiness of the institutions were the niggardliness of the city authorities, the dishonesty of the politicians and their henchmen, and the public's lack of interest. Those who ran the hospital and those who appropriated the funds wanted to keep expenditures low; they saw nothing to be gained by coddling the inmates, who had few votes and little influence, and they knew that the public would object to the consequent increase in taxes if the costs of the almshouse increased. Although not a major source of graft except when building contracts were awarded, hospitals offered many opportunities for petty extortion. Soon after the opening of the public almshouse in Philadelphia, complaints were made "against the Overseers of the Poor who had supplied the Poor with Necessaries out of their own Stores and Shops at exorbitant Prices, and also against Overseers who had paid unreasonable Accounts to their Friends and Dependents for Services done the Poor."[41]

Superintendents or wardens of almshouse hospitals were laymen whose main responsibilities were seen as controlling a crowd of vagrants, alcoholics, prostitutes, and other criminals as well as the insane, and making sure that they worked, when possible, to pay back some of what the taxpayers spent on them. Appointed by politicians, superintendents consorted with politicians and shared their habits. They had their own sources of petty graft, such as the sale of dead bodies for dissection.[42]

Thus, if we attempt to evaluate patient care in the almshouse hospitals from the few known facts, it appears that there were periods of reasonably adequate space and staff alernating with longer periods when both were scarce. Many nurses and attendants were inexperienced and uninterested. Food and supplies were probably barely sufficient in the best of times; with poor management, graft, overcrowding, or epidemics serious shortages existed. Whether patients were well or ill housed, poorly or adequately fed, properly or carelessly attended depended less on the decisions of experts and the informed consent of the public than on the relative influence of acquisitiveness versus conscientiousness among politicians and officeholders.

Another function of the almshouse hospitals was to educate medical students. Doctors were badly needed in the rapidly growing colonies, and at first the only way they could be taught was as apprentices. From the start, apprentices of attending physicians and surgeons were admitted to the wards of the Philadelphia almshouse. As the numbers of students grew, separate rooms were provided for them. Organized teaching probably began at the almshouse in 1770, when groups of students were allowed to observe patients in labor. This followed closely the opening in 1765 of the medical school of the College of Philadelphia (later the University of Pennsylvania). Other clinics and lectures were added, and students came to the hospital in increasing numbers. At first access to almshouse patients was granted as a privilege, but in 1805 students were required to purchases tickets. Not until 1860 were students again admitted without a fee, and the last vestige of the students' apprenticeship to the hospital disappeared; thereafter, the relationship of students to the hospital was through medical schools.[43]

The lay managers of the Philadelphia almshouse were not always as enthusiastic about teaching as the doctors. For nearly fifteen years, from 1789 to 1803, teaching was suspended altogether because the two groups could not agree. After it was started again, the number of students at the hospital increased irregularly from fifty-three in 1818 to 185 in 1830. After several stops and starts during the years of political control, permanent access was finally obtained in 1859.[44]

Students in almshouse hospitals were usually observers only, gaining close contact with patients in their preceptors' offices or in patients' homes. A fortunate few obtained positions as house officers. By 1802 there were at the Philadelphia almshouse a senior and a junior house pupil and an apothecary. The first served for two and a half years and the last two for three and a half years. In 1820, when their number had increased to eight, the title of house pupil was changed to house physician and house surgeon. They lived in the hospital but paid a fee, which at first was $80 and rose gradually to $250 by 1835. Beginning in 1839 the fee was reduced, and by 1866 house officers no longer paid a fee and were given board and room. That the institution could collect a fee instead of paying a salary was obviously a reflection of the student's need for close contact with patients and the low esteem in which his skills were held. By the end of the almshouse period students were better trained and of more value to the hospital, and some house officers were graduates instead of students.[45]

At Bellevue, as at the Philadelphia almshouse, teaching began early and at first depended on what individual attending physicians chose to do for their students. In 1787 Nicholas Romayne established a private medical school and utilized the almshouse "to teach clinical medicine for the first time in New York," but the school lasted only a few years. When the College of Physicians and Surgeons opened in 1807, it used the almshouse for teaching from the beginning. Students of New York University were also taught at Bellevue after its medical school opened in 1841. For the fifteen years preceding 1847, when the hospital was in the grip of the city politicians, teaching was at a low ebb, but the tide turned swiftly when an attending staff was again placed in control. In 1849 an amphitheater was opened, and the custom arose whereby staff physicians and surgeons gave lectures, which were attended by their own and other students. An added impetus was given to teaching of medical students shortly after 1861, when the Bellevue Hospital Medical College opened with a curriculum that was almost wholly based at Bellevue.[46]

Instruction of students at the Baltimore almshouse began around 1812, when two professors at the College of Medicine of Maryland (later the University of Maryland) began teaching there. By 1818 all seven attending physicians and surgeons were on the faculty of the new school, and students were allowed to accompany them on rounds by paying $10 a year. Two students were also selected to live at the hospital on payment of $150 a year.[47]

Thus, many city hospitals in the United States began as almshouses. By the time of the Civil War several of the older ones had either divorced themselves from the poorhouse or were taking over leadership of the almshouse-hospital complex. The concept of separating the general hospital from workhouse, old-age home, and insane asylum had been generally accepted as a proper objective, and the cities were in different stages of bringing this about. Medical care was generally poor as a result of the paucity of good doctors, the absence of trained nurses, the interference of politicians for their own ends, the indifference of the public, and the scarcity of funds for decent food and quarters and adequate help. Toward the end of the period, care improved somewhat as better resident physicians were appointed and medical students were better trained. In medical education the apprentice system of training had run its course, having been displaced by ward walks, lectures, and demonstrations by attending doctors who were on the faculty of a medical school. The city hospitals and the medical schools loosely affiliated with them were thus prepared to welcome the crowds of students who flocked to them after the war.

# Part II · *The Practitioner Period*

# 2 · Cities, Politicians, and City Hospitals

> I view great cities as pestilential to the morals, the health, and the liberties of man.
>
> Thomas Jefferson

> With very few exceptions, the city governments of the United States are the worst in Christendom—the most expensive, the most inefficient and the most corrupt.
>
> Andrew D. White

> A municipal hospital is an index of a city's regard for health. If it is a bad hospital, slovenly, dangerous, wasteful and graft-ridden, it is a measure of the city's intelligence and ideals. If it is a good hospital, it has been made and kept so...by faithful men and women fulfilling the desires and intents of the community.
>
> Major-General Hugh S. Cumming

CITIES CONTAINED ONLY a fraction of the total population when the United States became a nation: at the time of the first census, in 1790, a mere 5 percent of the inhabitants lived in communities of 2,500 or more. But the urban dwellers multiplied rapidly; in 1860 they made up 20 percent of the population and in the decade following 1910 became the majority. Large cities, even though Jefferson disparaged them, shared in this growth: in 1810 none contained as many as 100,000 people, but in 1860 nine exceeded this size, and in 1910 fifty.[1] Between 1860 and 1910 the populations of Baltimore, Boston, New Orleans, and Philadelphia increased from twofold to almost fourfold, while that of the new capital city of Washington increased fivefold (see Table 2). New York, which had experienced a phenomenal gain in the seventy years preceding 1860, continued to expand sixfold in the next fifty years. But the real population explosion was in Chicago, which grew from 100,000 to over two million in the fifty years after 1860.

Hospitals in 1860 still housed only a few of the sick; people who could manage it were cared for at home. But the poor often had no place to lie down at home, if they had a home, and no one to attend them there. The masses of tired and huddled poor increased daily in the larger cities, for this was the time of the great immigration

Table 2. Population of selected cities, 1790–1910.

| City | 1790 | 1860 | 1910 |
|------|------|------|------|
| Baltimore | 13,503 | 212,418 | 558,485 |
| Boston | 18,038 | 177,840 | 670,585 |
| Chicago | — | 109,260 | 2,185,283 |
| New Orleans | a | 168,675 | 339,075 |
| New York City | 32,305 | 813,669 | 4,766,883 |
| Philadelphia | 28,522 | 565,529 | 1,549,008 |
| Washington | — | 61,122* | 331,069 |

*Sources:* For 1790: U.S. Department of Commerce and Labor, *A Century of Population Growth from the First Census of the United States to the Twelfth, 1790–1900* (Washington, D.C.: U.S. Government Printing Office, 1909), p.78. For 1860: U.S. Census Office, *Statistics of the United States...in 1860, Compiled from the Original Returns...of the Eighth Census* (1866; reprint ed., New York: Arno Press, 1976), pp. lvii–lviii. For 1910: U.S. Department of Commerce, Bureau of the Census, *Statistical Abstract of the United States, 1941* (Washington, D.C.: U.S. Government Printing Office, 1942), pp. 27–29.

a. When Louisiana became part of the United States in 1803, New Orleans was a town of about eight thousand persons, more than half of them black. The 1810 census reported 17,242 inhabitants. Hodding Carter, William Ranson Hogan, John W. Lawrence, and Betty Wertlein Carter, eds., *The Past as Prelude: New Orleans, 1718–1968* (New Orleans: Tulane University, 1968), p. 37.

stampede: between 1850 and 1860 the number of foreign born in the United States almost doubled, from 2.2 to four million; in 1860 they made up 13 percent of the population and remained near that proportion for the next half century, while the total population tripled. Most of the immigrants from abroad settled in cities. In 1860 47 percent of the people in New York City were foreign born; in Chicago 50 percent; in Boston 36 percent; in Philadelphia 30 percent; and in Baltimore 25 percent.[2] The newcomers often arrived in poor health and were highly susceptible to illnesses. Their living conditions upon arrival, crowded and unsanitary, were hardly conducive to good health, and they had little knowledge, few friends, and no means by which to cope with illness. They were public charges, and the cities had to care for them as best they could.

A swelling population and more strangers within the gates increased the need for hospital beds. The demand was augmented further for a totally different reason: hospitalization became more desirable because doctors learned how to combat the spread of infection in the hospital and thus prevent many infections from developing. For centuries it had been known that some diseases were contagious.

Communities acted to prevent their spread by isolating the victims in pesthouses placed well apart from other buildings. The principal buildings of a hospital, however, were built on the block system, in which wards were used as thoroughfares to other wards. As people became convinced that not only the highly contagious diseases but other infections too were transmitted from person to person, they tried to limit their spread by building hospitals on the pavilion system, whereby wards were placed at the ends of corridors and kept separate from other wards. This architectural style began in the United States around 1860 and had run its course by 1930.[3] It may have helped prevent some infections, but many still passed from one patient to another and sometimes through an entire hospital.

More effective procedures were needed before infections could be substantially diminished. These were devised by Ignaz Philip Semmelweis of Vienna and Joseph Lister of Edinburgh. The former demonstrated in 1847 that puerperal fever (fever following childbirth) could frequently be prevented by cleanliness of doctors' and attendants' hands and objects in contact with the patient. Lister showed in 1865 that the spread of infections during and after operations could be controlled by the use of phenol, then called carbolic acid. Shortly thereafter this method, antisepsis, was replaced by asepsis, in which everything approaching the surgical field was sterilized or rendered scrupulously clean. The results of antisepsis and asepsis were dramatic: death rates after operations were drastically lowered; surgical procedures were extended into formerly forbidden areas of the body; infections no longer spread like wildfire through hospital wards; and patients admitted to the hospital now stood a good chance of coming out alive. Once these new concepts had been adopted by doctors and the public—and this took several decades—the hospitals' sphere was greatly widened. From being an object shunned, the hospital became a desirable place for sick people to go. The number of beds in government-owned hospitals increased from 11,767 in 1873 to 24,802 in 1889, a 211 percent increase. In the same period the number of beds in private hospitals increased by 283 percent, from 15,252 to 43,120.[4] The upper and middle classes sought the benefits of hospitalization, and the private hospitals concentrated on serving them.[5] The poor were left mostly to the public hospitals, especially in the large cities, where sizable populations lived on low-paying or intermittent jobs or had no work at all.

To meet the challenge, hospital units emerged from their poorhouse shells and evolved into general hospitals, such as Philadelphia General and Bellevue. A few cities met the need for general hospitals for the poor by establishing institutions independent of almshouses. Notable examples were Cook County Hospital in

Chicago and the Boston City Hospital. Both were built by the 1860s, and both became prominent institutions by the end of the nineteenth century.

Chicago had been incorporated as a city in 1837. The population boom, which started some years before that and which was hurried along by successive waves of immigration beginning in the 1830s, was only temporarily halted by the Great Fire of 1871. The city began rebuilding immediately. It continued to stuff into its boxlike houses the thousands of immigrants who came to build its streets and homes, to man its railroads and meat plants, to beg in its gutters, and to die in its hospitals. In 1843 the city's first hospital was built, a shack of rough boards for patients with contagious diseases; and in 1845, or shortly before, an almshouse was erected in the township of Jefferson, later Norwood Park, twelve miles northeast of downtown Chicago. Because these did not suffice for the teaching of medical students, several professors from Rush Medical College, which had been organized in 1843, helped to found a private hospital. This opened in 1850 as the Illinois General Hospital of the Lakes, a title almost larger than its twelve-bed capacity. In the following year the hospital was taken over by the Sisters of Mercy and soon thereafter was named Mercy Hospital, which it is still called today.[6]

In 1854 the city itself built a small frame hospital two miles south of the business center, followed three years later by a larger one at the same site. This remained idle until 1859 while the regular medical profession disputed with the homeopaths over who would staff it. (The latter were members of a medical sect who claimed to cure diseases with minute doses of drugs that produced symptoms similar to those caused by the diseases themselves.) To evade the problem, city authorities leased the hospital to doctors representing Rush Medical College, who agreed to care for the city's patients at three dollars per patient per week. But in 1862 the federal government commandeered the building for use as an army hospital during the Civil War. After the war the city leased it to the county, and it opened as Cook County Hospital in January 1866, which marked the beginning of continuous service in a general hospital controlled by the county.[7]

Up to this time county authorities had shifted the indigent sick back and forth—like boxcars in Chicago's sprawling railroad yards—wherever expediency directed: first to the almshouse; then in 1847 to a rented warehouse called Tippecanoe Hall; from 1851 to 1863 to Mercy Hospital (at three dollars per week); and then to the almshouse again.[8] Finally, in 1866, they were transferred to the newly opened Cook County Hospital. It was an imposing structure of brick and limestone, containing 130 beds, to which a frame wing added ninety more in 1870. Only Mercy Hospital exceeded it in size.[9] William E. Quine, an early house officer and later dean of the University of Illinois medical school, recalled that the hospital was modern, well equipped, and liberally supported by the county, and that the "general atmosphere was such as surrounds a happy family."[10] This paradise, if paradise it was, bloomed but a short time; the Great Fire of 1871 consumed the wards completely. But the need for the hospital had been demonstrated, and, after much pressure from the medical staff, a new hospital of 350 beds was built two miles west of the business district on the site of the present Cook County Hospital. This opened in 1876.[11] The new hospital grew and became well known in the Midwest and later throughout the nation. In 1900 it had a capacity of 1,100 beds and admitted 22,460 patients. Between its opening in 1876 and the end of the century, four more buildings were added.[12]

At the time of Chicago's incorporation Boston had existed for over two hundred years; for most of that time it had retained a homogeneous population of descendants of the original Anglo-Saxon settlers. In 1830 only 5.6 percent of the city's sixty thousand people were foreign born. But the old-stock Yankees were soon to be confronted with a "green wave," as Irish landlords began to force tenants off their land early in the century and the potato famines, which began in 1845, sent one-fourth of Ireland's population to America. By 1860 36 percent of Boston's population was foreign born, and almost three-quarters of the newcomers were Irish.[13]

The almshouse and the pesthouse had offered refuge for those who could not be nursed in their homes until the Massachusetts General Hospital opened in 1821. The public had been assured that this institution, which was controlled by a board of private citizens, would care for the city's sick poor and would relieve the almshouse of that responsibility. But the hospital soon found itself swamped with patients, largely immigrants. In 1849 the city council pointed out that more than one-quarter of the applicants to the Massachusetts General Hospital had been refused admission in recent years "either on account of lack of accommodations or for some other reason."[14] One other reason was that the hospital's trustees had resisted admitting Irish patients except in emergencies, claiming that "admission of such patients creates in the minds of our citizens a prejudice against the Hospital, making them unwilling to enter it,—and thus tends directly to lower the general standing and character of its inmates." Later, "moved by a sense of duty and humanity," the hospital opened its wards to all nationalities, and by 1865 foreign patients outnumbered native Bostonians.[15]

The Boston Lying-In Hospital, another private hospital, had opened in 1832 but had been forced to close for lack of funds in 1857 and did not reopen until 1873. This left Boston, with a population in 1860 of nearly 178,000, with only two hospitals: Massachusetts General, with 140 beds, and Carney, a small hospital of thirty beds owned and run by the Sisters of Charity.[16]

The Massachusetts General Hospital indirectly helped the cause of those who proposed a city hospital. It had existed from the beginning on gifts, bequests, and the small fees the semi-indigent could pay. The advocates of a city hospital naïvely anticipated that it would cost the taxpayers little or nothing because gifts and fees would pay its running expenses.[17] Although this was not to be, the concept helped emphasize that the city hospital was not an almshouse hospital and that it could be expected to function similarly to the Massachusetts General. As it turned out, although the city hospital, like public hospitals elsewhere, had to be sustained mainly on tax revenues, it and the Massachusetts General often charted similar courses.

A movement to build a permanent city hospital had been gathering force ever since the city had successfully operated a hospital during the cholera epidemic of 1849. Plans for a hospital in the western part of the city, where blocks of fashionable houses were being built on land formed by filling in the Back Bay, were thwarted by local citizens. But the campaign, once started, could not be stopped, and the city fathers made use of a bequest received by the city some years before. The successful removal of a urinary calculus by John Collins Warren at the Massachusetts General Hospital had moved Elisha Goodnow to repay his debt to the doctors by leaving the bulk of his estate to the city for a hospital, which he had specified should be built in the less fashionable South End. His bequest of $26,000 was supplemented with city funds, and in 1858 the General Court of Massachusetts authorized the city to establish and maintain "a hospital for the reception of persons who by misfortune or poverty may require relief during temporary sickness." Construction was started in 1861 at the present site in the South End.[18]

Boston went about building its city hospital more thoughtfully and methodically than Chicago. The city council offered a prize for the best plan, which was awarded to Dr. John Green.[19] In his treatise the winner recognized especially the danger of infections:

> Epidemic and contagious diseases often arise in large hospitals, and have several times, in our own city, rendered it necessary to close the Massachusetts General Hospital against all applicants for admission. Many of the most celebrated hospitals are familiar with these terrible visitations; and, in others, the uniformly bad condition of all open sores testifies but too truly to the wretched sanitary condition of the wards.[20]

But, he contended, "epidemic diseases are not...the necessary condition of the association of the sick" and recommended: excluding patients with contagious diseases from the general wards; arranging hospital buildings to provide "the perfect isolation of every ward from every other ward"—in other words, the pavilion system; and locating wards so as to admit as much sunlight as possible.[21] Given the imperfect state of knowledge at the time, this was about as far as hospital construction could go in preventing infections. Whether in response to the prize essay or not, Green's methods were followed in building the Boston City Hospital.

Completed in 1864, the hospital consisted of medical and surgical pavilions containing a total of 225 beds and a central administration building with an impressive dome that rose 148 feet above the street. But if the latter seemed a bit grandiose, the purpose of the hospital was utilitarian enough, as expressed in Mayor Alexander H. Rice's appeal for its founding. It was to be "an asylum for the industrious and honest mechanic and laborer who by sudden injury or disease is temporarily prevented from laboring for the support of himself and family" and "a home to which the respectable domestic may be sent...whose attic chambers cannot be made comfortable, and who cannot receive the requisite attendance."[22] In other words, this was to be a hospital for the deserving poor, not an almshouse hospital. Although this hospital found that the poor could not be separated into strata and that it was required to accept all comers, its high objectives tended to elevate the staff's morale and the public's image of the institution.

Another constant spur was the close relationship maintained with the elder brother, the Massachusetts General Hospital, which included a healthy amount of sibling rivalry. Sometimes the Boston City was first to try a new procedure, such as the use of phenol in September 1867 (see Chapter 6), followed by the Massachusetts General one month later. Often the earlier established hospital was the first to try something new; for instance, it began an ambulance service in 1873, followed four years later by the Boston City. The city hospital, recognizing that one of its important functions was to care for victims of accidents, opened in 1902 a twenty-bed branch hospital in the commercial district, where the accident toll was heavy. Thus, this hospital perceived early that it should carry its services to the people, a lesson some city hospitals have still not learned.[23]

Except that they were still connected with almshouses, the public hospitals in many of the large cities of the United States resembled the Boston City and Cook County Hospitals around 1860. The Philadelphia Hospital, although separated in name and function from the rest of the almshouse complex, was still administratively tied to the remainder and inevitably influenced by the prevailing philosophy of long-term custodial (do-nothing) care. The average census of the hospital and almshouse for the year ending July 1, 1860, was 2,520 and for the insane department 425; 6,176 patients were treated on the adult and children's wards. The hospital was divided into three departments—medical, surgical, and obstetrical—each staffed by four doctors. By 1877 there were eight for each. Six wooden pavilions for medical patients were added in 1875 and a maternity pavilion in 1883–1884.[24]

By 1860 Bellevue had shed the burden of the almshouse and workhouse. Twelve years earlier the last of the prisoners and indigents had been moved away, and the hospital moved into a massive building vacated by the almshouse. The building was further enlarged in the mid-1850s by the addition of a fourth story and a wing. The structural improvements and additions that equipped Bellevue for its new role as a general hospital for acutely ill patients increased its bed capacity to 1,200 and enabled its staff to treat 11,411 patients in 1860 and increasing numbers in subsequent years. Its facilities were sorely needed because only one other hospital had been built between its opening and the mid-nineteenth century, the New York Hospital, which opened in the 1790s. In the decade beginning in 1850, twelve new hospitals had been established in Manhattan.[25] Bellevue's focus on a single goal, the care of the acutely ill, sharpened its view of itself and the community to which it ministered. As a result, during the next two decades it pioneered in several areas: the better to serve the community it established an outpatient department and an ambulance service; the better to serve its patients it founded a nurses' training school; the better to serve medical education it placed each medical school in charge of its own medical and surgical service.[26] Many considered it to be the leading city hospital in the country during the half century before 1900.[27]

During the period in which the public hospitals of most cities were freeing themselves of other responsibilities, those in Baltimore and Washington continued to be dominated by the philosophy of the almshouse and workhouse. In Baltimore the patients and inmates of the almshouse complex at Calverton were transferred to the Bay View Asylum when it opened in 1866. The huge building (714 feet long) set high on a hill afforded a beautiful view of the harbor and the Chesapeake Bay, but its objectives were no more forward looking than those of its predecessor. It continued to be a catchall for the petty thief, the vagrant, the penniless widow, the worn-out laborer, the insane, and the chronic invalid, who were sent there by the various agencies of the city, and the patients the other hospitals didn't want.[28] At first all were housed in the same building, and the same administration governed all units, even after the psychiatric and tuberculous patients were moved into separate buildings in the eighties and nineties. Although the medical staff did their best to care for acutely ill patients in wards set aside in the main building, the institution was primarily an almshouse. In 1890, for instance, nearly 80 percent of the inmates were listed as permanent residents.[29]

Unlike the other cities, Washington had several civilian hospitals that were built partly or wholly with funds from the federal government. Sometime during the Civil War a hospital was opened as an emergency facility for destitute blacks, who crowded into the city by the thousands, assuming that a nation that fought for them would care for them. This institution became Freedmen's Hospital and was for many years supported, though grudgingly and inadequately, by the federal government and at times by the District of Columbia as well. Although Freedmen's accepted paying patients, many could pay little or nothing.[30]

Two private hospitals were partially funded by the federal government for the care of Washington's underprivileged or indigent patients. Providence Hospital was opened by the Sisters of Charity in 1861, and up to 1929 the surgeon general of the army contracted annually with them for the care of the nonresident poor. Columbia Hospital for Women, founded in 1866, included representatives of the federal government on its board of trustees and received federal funds for the support of free patients. By arrangement with the Secretary of War, the hospital set aside twenty beds for wives and widows of soldiers and sailors.[31]

With three hospitals receiving support from the federal and local governments, Congress, which appropriated all funds for the District of Columbia (including those derived from local taxes), saw no reason for a general hospital operated by the city government. Yet, willy-nilly, that is precisely what was evolving at the Washington Asylum. The three-story brick building built in 1859 housed petty criminals, almshouse and insane patients, and some persons who were acutely ill. After the Civil War the hospital facilities were augmented by incorporating several one- and two-story frame buildings, which had contained single wards of thirty beds when they were used as a military hospital. By 1897, when outside experts inspected the hospital, they found that it consisted entirely of frame buildings and concluded that the "standard of care and provision of the sick [was] but little above that of the primitive country poorhouse of an earlier day." The hospital functions were distinctly subordinate to the almshouse functions and to the affairs of the city jail. The sick were taken care of by a resident physician and a few medical students under the supervision of a medical director, who was in attendance a few hours each day. Thus, at the beginning of the twentieth century, except for the introduction of some new techniques such as anesthesia and asepsis, the situation and practices of the publicly controlled hospital of the nation's capital were similar to those of other eastern cities a half century before.[32]

Charity Hospital of New Orleans differed from other government-owned hospitals for the poor in American cities: it originated as a private hospital and later was taken over by the state; thus, it functioned as the city hospital for New Orleans. Founded in 1736 by a bequest from a French sailor, Jean Louis, it survived precariously through hurricanes, fire, and a succession of governments: first under an appointed board, then under the government and the church during Spanish rule, and then under the patronage of a wealthy nobleman, Don Andres Almonester y Roxas. Finally, in 1811 the legislature of the new territory of Louisiana, in order to settle a continuing dispute between the city of New Orleans and the Almonester heirs, declared the hospital in effect public property and placed it under a board of six citizens appointed by the governor and three members of the New Orleans city council.[33]

While intended solely for the care of the sick, Charity Hospital also functioned for a long time as an almshouse for lack of other facilities for the able-bodied poor. In 1818 a clergyman wrote that "patients of every description...were promiscuously blended together, and doomed alike to such unfailing neglect as often resulted in untimely deaths." In January 1848 the hospital, built to accommodate 540, had over eight hundred patients on its wards. Annual pleas to the legislature for more facilities were not effective. By 1860 the hospital was treating an average of twelve thousand patients each year.[34] Crowding was intensified after the Civil War by an overload of white patients impoverished by the war and ex-slaves who were no longer the responsibility of their former owners. By 1879

> the desolation and want...was appalling. Beds and bedding, provisions, drugs, and necessaries for the patients were sadly deficient, or totally exhausted, the beds not having a change of sheeting, and scarcely any winter covering, the supply of drugs and provisions being totally exhausted, while the credit of the Hospital was gone, and could not be used for the purchase of a pound of flour or salt.
>
> The buildings, too, [were] in a most dilapidated condition, the floors, walls and roofs being unsafe. In places, the walls and roofing had already succumbed...; other portions were threatening and in danger of falling in;...and the supply of water (that most necessary element in a hospital), was very limited.[35]

Soon a new board of managers, supported by a forceful governor, began rebuilding the hospital. After urgent repairs were made and some basic necessities supplied (ranging from cold and hot water for the operating room to the separation of children from adults), the first of the sorely needed buildings could be erected. By the end of the century the hospital had caught up with comparable institutions in the North. For example, in 1894 a school for nurses was started (see Chapter 4), and an ambulance service began the following year.[36]

As the cities of the Midwest and Far West had grown, they too had needed to provide for their sick poor, and many of them chose to combine the hospital with the almshouse. The march of urban civilization across the continent is reflected in the dates when cities established almshouse-hospital complexes: Cincinnati in 1823, Cleveland around 1837, Denver in 1862, Seattle in 1877, and Los Angeles in 1878. During the second half of the nineteenth century these hospitals for the most part shook loose from the almshouses, workhouses, and jails and became general hospitals for the care of acute illnesses.[37]

Difficult and complex as their problems were, the public hospitals and almshouses were not free to solve them on their own. Rather, they were caught up in the destinies of the crowded, bustling, frenetic anthills that American cities became after the Civil War. Within the confines of these cities "were focused all the new economic forces: the vast accumulations of capital, the business and financial institutions, the spreading of railway yards, the gaunt, smoky mills, the white-collar middle classes, the motley wage-earning population."[38] Here could be found concentrated wealth and concentrated poverty. The dilemma of the city government was how to use the first to help the second. For certain individuals the dilemma offered the opportunity to use the wealth to line their own pockets.

City governments tended to copy the federal model, with a chief executive—the mayor—and two legislative bodies, but here the resemblance ended. Many necessary governmental powers were retained by state legislatures. The mayor's powers were sharply curtailed by the legislative councils, the members of which retained office by trading jobs and votes. As cities expanded rapidly after the Civil War, they stretched the capacities of their public facilities to and beyond their limits. When an important job had to be done—the building of a new bridge or a city hall, the administration of an orphan's home or the city water works—it was put in the hands of an independent board or commission. For instance, by 1890 Chicago had eleven major governmental units, each an independent fiefdom with taxing and regulatory powers.[39]

In keeping with the British poor laws, American cities usually lumped the sick together with others who needed public aid: the poor, the aged, orphans, the insane, and the chronically disabled. Governance of local public hospitals in the nineteenth century was characteristically vested in a board of citizens, elected or appointed and responsible to the mayor or chief executive officer of the political unit

35

(see Table 3). Sometimes this board controlled only the almshouse-hospital complex, as at Bellevue preceding 1860, and sometimes the city's welfare services also, as in New York from 1860 to 1902 and Baltimore for a century beginning in 1866. Sometimes the control of public safety was added, as in New York from 1860 to 1896 and in Baltimore from 1900 to 1935. These diverse responsibilities often allowed the size and inertia of the custodial operation to smother the dynamics of the hospital function. Once it was realized that combining all these services under one administration could be detrimental to all, hospitals were placed in separate units of government. The Boston City Hospital had its own board of trustees from the start, and Bellevue was removed from the Department of Public Charities in 1902 and placed under a separate governing board along with several other city hospitals.[40] At Cook County Hospital, in contrast, control remained directly under the county commissioners until 1970, a situation that greatly increased the politicization, administrative confusion, and insensitivity to proper medical care characteristic of that hospital in the twentieth century.

Politics affected the boards of citizens too. In 1880 the state legislature was urged to incorporate the board of trustees of the Boston City Hospital rather than continue appointments for one-year terms by the city council. Incorporation was recommended because the proposed five-year staggered terms would make for stability and give trustees more opportunity to become familiar with the hospital's problems, and it was anticipated that the five trustees appointed by the mayor would be freer from political entanglements than trustees appointed by the city council and that they would be prominent citizens who would bring prestige to the hospital and attract gifts. On the contrary, a minority of the governing board wanted to retain the existing system, claiming that it represented more directly the persons who were receiving the hospital's services.[41]

The legislature made the changes the majority requested and in 1885 even eliminated the two positions for city councilmen. The board's partial insulation from political influence did attract many superior trustees and enabled them to focus on the general good of the hospital, but the new system did not banish politics. As the mayor had to be responsive to his electorate, so did his appointees, especially since they had to go hat in hand for their appropriations. The city council still exerted an influence, even though it was now indirect, and as immigrant-based machines took control of Boston's government, the Brahmin elite were eventually displaced from the board. The machines tried to use the hospital for their purposes, sometimes helping and sometimes harming the patients in the process.[42]

**Table 3. Changes in governance of selected public-general hospitals (from date of establishment at present site).**

| Period | Governing body |
|---|---|
| *Bay View Asylum—Baltimore City Hospitals* | |
| 1866–1899 | Trustees of the Poor |
| 1900–1935 | Department of Charities and Correction, Board of Supervisors of City Charities |
| 1935–1965 | Department of Public Welfare |
| 1965– | Department of Hospitals |
| *Bellevue Hospital, New York City* | |
| 1816–1849 | Board of Commissioners of the Almshouse |
| 1849–1860 | Board of Governors of the Almshouse Department |
| 1860–1896 | Board of Commissioners of Public Charities and Correction |
| 1896–1902 | Department of Public Charities |
| 1902–1929 | Trustees of Bellevue and Allied Hospitals |
| 1929–1970 | Department of Hospitals |
| 1970– | New York City Health and Hospitals Corporation |
| *Boston City Hospital* | |
| 1864–1966 | Board of Trustees of the Boston City Hospital, Department of Hospitals |
| 1966– | Department of Health and Hospitals, Division of Hospital Services |
| *Cook County Hospital, Chicago* | |
| 1866–1871 | Board of Supervisors of Cook County |
| 1871–1970 | Board of Commissioners of Cook County |
| 1970–1979 | Health and Hospitals Governing Commission |
| 1979– | Board of Commissioners of Cook County |
| *Washington Asylum Hospital—Gallinger Municipal Hospital— District of Columbia General Hospital* | |
| 1839–1871 | City of Washington |
| 1871–1890 | Board of Health |
| 1890–1900 | Superintendent of Charities |
| 1900–1926 | Board of Charities |
| 1926–1937 | Board of Public Welfare |
| 1937–1953 | Health Department |
| 1953–1970 | Department of Public Health |
| 1970–1977 | Department of Human Resources, Community Health and Hospitals Division |
| 1977– | D.C. General Hospital Commission |

Even if the citizens' boards in every city had been filled with honest, knowledgeable, dedicated people who had time to spend on the multitudinous concerns of the office—which they seldom were—they were hamstrung by perennially inadequate funds, buildings that were continually being outgrown, and a public philosophy that resented the cost of even the inadequate facilities. In an age in which it was held that hard work would bring success, failure had to be caused by not trying hard enough. Why, then, should the worker pay for the slacker?

The recognition of the contagious nature of some diseases also created added responsibilities for governments. The control of infectious diseases and the management of quarantine stations and isolation hospitals were usually placed under a separate board of health, which made sense as far as city administration was concerned but which fragmented the care of the sick. The multitude of officials and agencies concerned with health affairs often left the individual citizen bewildered. Jane Addams, working in a settlement house in Chicago, observed that "the hospitals, the county agencies, and State asylums are often but vague rumors to the people who need them most." The dilemmas of the native citizen were bad enough; those of the newcomer were infinitely worse. Along with the waves of new arrivals from Europe were substantial numbers of migrants from rural areas in the United States, especially blacks in border and southern cities. Needy, unsure citizens, fragmented governments, and a public for the most part uninterested all provided opportunities for the politician. Too often the sick person or his family, feeling unable to approach a government agency, turned instead to the precinct captain, the representative of what the insider called the organization and the outsider called the machine.[43]

The machine was an extralegal political organization that won elections and controlled key public offices and agencies. It existed because the size and complexity of the city made its government unwieldy and impersonal. Controlling the machines and making them personal for many citizens were the bosses and their henchmen. These men collected and shared the graft, which was abundant, while pushing a few causes in favor of the poor, providing jobs for the faithful, and doling out favors: a basket of food at Christmas for the widow with three children, a scuttle of coal when the father of the family was out of work, a word to the judge when Johnny got into trouble, and a bed at the hospital for those who didn't know how to go about getting it on their own.[44]

Hospitals were not usually in the center of the politician's orbit. For one thing, the average ward heeler had scant understanding of and less interest in the technical achievements of medicine; for another he had little traffic with the doctors, who were either in the upper classes or striving mightily to get there. Nor were the city hospitals prolific sources of graft. Little return could be expected from the unfortunate patients and their families except votes at the next election. Most of the hospital jobs available to the politician as patronage were at the lower end of the scale. The higher positions demanded more education than would be found among most of the supporters of the machine.

At times, however, a machine managed to involve itself closely in the affairs of a hospital and to do quite well for itself. At Cook County Hospital in the 1880s the going price a doctor paid to a county commissioner for appointment to the attending staff (which brought the doctor no pay, only experience and prestige) was $1,000. In Philadelphia doctors in the almshouse hospital noted with irony that while their patients went hungry and cold because food was scanty and blankets were lacking, the political boss's yacht was moored at the almshouse wharf "to be stocked with coal, provisions, liquors, and cigars, preliminary to a wild, bacchanalian cruise." Small wonder that doctors nicknamed the hospital's governing board the "Board of Buzzards."[45]

A similar incident occurred in Cook County. In 1880 a gambling-house proprietor, Harry Varnell, became warden of the Cook County insane asylum and "promptly transformed the place into a country club for politicians...parties and dances were held almost every night, and rare foods and wines were served." In this case somebody snitched, and Varnell served a year in the penitentiary for his misdeeds; but soon after his release his gambling-house business was thriving again.[46]

Excesses like these disgusted doctors and the public alike on the few occasions when the searchlight of the law revealed them and the newspapers made copy of them. Most of the time they did not come to the public's attention, but the continuous succession of errors, stupidities, and villanies of politicians, as a leading member of the staff of the Philadelphia almshouse put it, wore down the morale of doctors, nurses, and attendants at the hospital, who saw the needs of patients and were trying to respond to them.[47]

At first Boston City Hospital was run by the Brahmins of the city for what they viewed as the benefit of the lower classes. But in the 1880s, with the election of Hugh O'Brien as the first mayor not born in this country, the Irish began to dominate the politics of the city. Their machines, run by such masters of the art as Patrick Collins, John F. ("Honey Fitz") Fitzgerald, Martin Lomasney, and, beginning in 1914 for the ensuing half century, James Michael Curley, encompassed the workings of the city hospital as well as the rest of the city government. Despite incorporation of the board of trustees, the hospital remained politically minded. The members found it expedient to grant favors that violated the rules they themselves had enacted, for instance issuing passes for visitors outside the official hours. The various ethnic groups among the employees were given their particular holidays off with pay, and, more important, nonprofessional jobs were filled by patronage. Occasionally, trustees appear to have influenced the appointment of a doctor to the attending staff, and at least one physician reached the staff by becoming a trustee first. This was anathema to most staff doctors, who subscribed to the principle that appointments should be made on merit alone, although they forgot how often belonging to the right family, graduating from the right medical school, and having the right friends were a necessary concomitant of merit and sometimes a substitute for it.[48]

Although the attitudes of doctors and politicians were usually worlds apart, they sometimes coincided. In 1903, when the central labor union of Boston asked for direct representation by labor on the board of trustees, claiming that labor represented the "chief patrons" of the hospital, it was turned down, and the *Boston Medical and Surgical Journal* sided with the trustees, commenting that "the fallacy and possible danger of such a principle of representation is apparent."[49] It would take another seventy years of troubled times in city hospitals before this stone wall of elitism began to crack.

Despite the incursions of politicians, the Boston City Hospital appears to have been freer from political influence than many similar hospitals during this period, although the editor of the *Boston Medical and Surgical Journal* probably went too far in 1910, when he claimed that "in spite of its relationship to the city government [the hospital] remains essentially without the sphere of politics." Some of the political influence seems to have had beneficial results in counteracting the tendency of the doctors to focus almost exclusively upon their professional duties and in bringing about a few changes that benefited the patient, such as liberalizing regulations for visitors, arranging longer hours for outpatients, and providing some of the pressure to open a neighborhood emergency station. A blend of politics and professionalism may have been the reason why the Boston City Hospital was considered one of the best hospitals in the United States by the turn of the century.[50]

By this time professionalism was beginning to affect the vocation of hospital administrator. The Association of Hospital Superintendents was established in 1899 to further the concept of a single, experienced administrator to be in charge of all aspects of the hospital's work while leaving decisions on medical affairs to the doctors. Some city hospitals had already moved in this direction. Among these was the Boston City Hospital, where two outstanding superintendents held office in the late nineteenth century—Edward Cowles from 1872 to 1879 and George H. M. Rowe from 1879 to 1909. Both were physicians who made careers of hospital administration and who gained national stature in the field.[51]

In contrast was the tawdry history of administration at Cook County Hospital during the same era. Because there was no hospital board to act as a buffer, the Cook County commissioners treated the hospital as they did any other branch of the local government: as a source of patronage and graft. Their appointees to the position of warden, the chief administrative officer, were laymen with little or no knowledge of hospital matters. The result, according to a former intern who became a distinguished surgeon, was that Cook County Hospital was "the most modern and the best built and worst managed hospital in the West." Not only was the administration of the hospital inept and corrupt; it was unstable. From 1866 to 1871 three wardens held office.[52] Expenditures for liquor in 1876 were almost double those for clothing and bedding, holders of patronage jobs were paid salaries out of proportion to their positions, and the hospital was labeled an "almshouse for the scum of the party ... beaten in ... elections." From 1876 to 1887 the affairs of the hospital were mismanaged by four especially corrupt wardens, one of whom was assisted by thirteen relatives whom he had put on the payroll. The fourth and apparently the worst of these was indicted for malfeasance, along with the hospital's chief engineer. It was claimed that the two had stolen $750,000 from contracts for coal, surgical instruments, and other supplies for the hospital and had divided this in the warden's office with ten county commissioners (all of whom were popularly called the Cook County boodle ring). The warden was convicted and sent to jail, and the politicians took cover for a few years.[53]

The next warden tried to impress the public with his concern for the taxpayer by saving money at the patients' expense. He badgered his subordinates to save the pennies while his superiors stole the dollars. For instance, he forbade interns to dress wounds without an order from the chief clerk, a layman, who decided whether the case justified the issuance of the necessary amount of antiseptic gauze. During the same period of parsimony, one of Christian Fenger's patients died from peritonitis when poor catgut used as a tie gave way and allowed feces to leak into the peritoneal cavity. When he saw the proof of this at autopsy, the renowned surgeon shook his fist in the direction of the warden's office. "Reformers? No!" he shouted. "No! Murderers!" But his anger stemmed in large part from awareness of his impotence.[54] (See Chapter 3 for a discussion of the contributions of Christian Fenger.)

Administration at the Philadelphia almshouse and hospital had its share of ups and downs. The doctor in charge of medical affairs, called warden at first and, after 1885, resident-in-chief, functioned most of the time under the supervision of the medical board, but at intervals the politicians abolished that board and appointed residents-in-chief responsible only to them. At such times this officer was likely to be a political hack, whereas under the medical board he was apparently an industrious and conscientious doctor with considerable administrative ability. The chief administrative officer of the hospital was called the steward up to 1873 and administrator later. As long as control of the hospital was under a board of guardians, he was a political appointee primarily interested in pleasing the politically minded board. In 1887 the almshouse and the hospital were placed under the Department of Charities and Correction of the city. From that time on the superintendents tended to be more conscientious and to cooperate closely with the medical board. Efficiency in the hospital improved further after 1903, when it was placed under a consolidated Department of Public Health and Charities and the house of correction was finally transferred to another department. An experienced hospital administrator was first placed in charge in 1920.[55]

In Baltimore and Washington the administration of the almshouse hospitals followed the same sequence as in the public hospitals of the larger cities, although some years behind them. The Bay View Asylum of Baltimore was run by a board called the Trustees of the Poor until 1900, when it was placed under a newly created Board of Supervisors of City Charities, which was responsible for all charity work in the city.[56] Lay superintendents, who were inexperienced in hospital matters, were appointed by these boards. The first professional administrator did not appear until the mid-thirties (see Chapter 5). In Washington administrative directions shifted according to the winds blowing through Congress at the time. In 1874, after a disastrous experience with local government resulting from the ineptitude of the governor appointed by President Grant, Congress established a board of three commissioners to run the city while they themselves made all the laws. In 1890 the institutions for the indigent were placed under a Superintendent of Charities, but in 1900 the structure was changed to a Board of Charities composed of five residents of the District of Columbia appointed by the President. This board immediately began pushing for separation of the hospital from the almshouse-jail complex and for a sorely needed general hospital building.[57] But a new hospital and professional administration were not achieved for almost thirty years (see Chapter 5).

It is apparent that two vastly different forces were affecting the management of city hospitals in the half century following the Civil War: politics and professional administration. Politics was the stronger and the dominant force during the first part of the period, but toward the end the growing professionalism of administrators began to make itself felt. A balance sheet showing the effects of political control would contain a few pluses and many minuses. Some historians in recent years have stressed the remoteness and depersonalization of local governments and emphasized the importance of political machines in making governments more personal, accessible, and responsive to the immediate needs of persons in the lower social classes—helping the underdogs to survive.[58] Granted that machines responded some of the time to some of the needs of some of the people, administering government so as to render justice to some who would otherwise be overlooked or harmed, yet in the collective sense the machines exacted an enormous price for these services. In hospitals particularly, the politicians' favors meant little. Most of the important decisions were made by experts and on the basis of objective criteria. As the knowledge and prestige of doctors increased, and as the training of house officers improved, the slip of paper that the ward heeler sent with the patient was likely to be thrown in the wastebasket, while the admission or nonadmission of the patient was determined by the results of a doctor's examination. Once the patient was in the hospital, the ward heeler, if he visited the patient at all, could only pretend to influence the treatment. Changes initiated by politicians in the regulations of a city hospital that were designed to benefit patients or their relatives were minor gains compared with the serious deficiencies in vital diagnostic and treatment services created by politicians who appointed incompetent and careless officials and attendants and stole the money that could have provided better food, more comfortable wards, sufficient bedclothing, and needed medicine and equipment.

The attitude of the medical staff toward the activities of politicians in hospitals was a curious mixture. Practically all doctors condemned the system, but usually they only grumbled among themselves over the shortages of help, supplies, and equipment. They remained exclusive and concentrated on professional matters, seldom taking the time to work with the politicians to improve the hospital. David Riesman was jolted one day when a health official reminded him that the medical profession of Philadelphia had not protested against the disgraceful conditions at the Philadelphia Hospital. When Riesman called this to the attention of the medical board, it prepared a list of needed improvements and got the support of the medical society, top political leaders, and the public for the city to spend $10 million for new buildings and renovations.[59] Furthermore, doctors did not hesitate to use politicians for personal ends. Doctors had to be willing to pay for appointments to Cook County Hospital or staff positions could not have been sold. Lawrence F. Flick, a high-principled intern at the Philadelphia Hospital in 1879, failed to see any similarity between his appointment upon the recommendation of a state senator and the city's political boss and the irregularity of a clerk's being paid for two jobs while performing one.[60]

In the nineteenth century the practices of the politicians were cruder than in later years, their thievery more obvious, and their plundering more extreme, and the public was more powerless than later. At the turn of the century James Bryce called the municipal governments the "one conspicuous failure of the United States" because of "the extravagance, corruption, and mismanagement [that marked] the administrations of most of the great cities."[61] City hospitals suffered much and profited little from these governments, yet their facilities grew and their services improved despite these handicaps.

# 3 · *Practitioner-Doctors, City Hospitals, and Medical Schools*

He was a verray, parfit practisour...And
gladly wolde he lerne, and gladly teche.
                    Chaucer, *Canterbury Tales*

Students were raw; material plenty; teachers
untrained. Didactic instruction was prominent; almost exclusive.
Description of Boston City Hospital, 1864

FORTUNATELY FOR THE sick poor, city hospitals
attracted many excellent doctors to their staffs. Both the newly
founded city hospitals and those that emerged from almshouses
offered the opportunity to see patients with a wide variety of illnesses,
enabling the ambitious neophyte to learn and the seasoned veteran to
keep his knowledge current and his skills sharpened. Moreover, staff
membership in a large city hospital conveyed distinction and status to
a select group in an era when a doctor might hang out a shingle
whether he had graduated from a leading medical school and spent
several years in advanced training or had received a diploma after a
brief correspondence course.

Staff doctors were practitioners. Their laborious days and nights
were crammed with office visits and house calls. Yet they found time
to pack in additional hours attending patients at the city hospital. The
pursuit of prestige alone could not explain this sacrifice; most of them
were driven by an age-old tradition that a doctor uses his best
knowledge and employs his keenest skills for each patient under his
care, whoever and wherever he might be. A prime example was David
Williams Cheever, one of the thirteen members of the attending staff
of the Boston City Hospital when it opened. His devotion to the city
hospital was such that he "considered that his hospital patients had
prior claim upon his time and attention. He could turn his private
patients over to an assistant, but...he and he alone was responsible"
for the sixty to one hundred city hospital patients under his care. He
remained on the staff for thirty-three years, and throughout that time
he taught Harvard medical students, becoming professor and head of
the school's department of surgery in 1882. Busy practitioners like
Cheever had scant time for research. But since new techniques and
new operations are often worked out on the cadaver before being
tried on patients, and since Cheever had access to cadavers as a

45

demonstrator in anatomy at Harvard, he was able to perform "many, bold, original and unusual operations that gave him an international reputation," according to one of his associates. He inaugurated surgical operating clinics and Sunday conferences for the hospital's surgical staff. He also found time to initiate and edit the *Medical and Surgical Reports of the Boston City Hospital*. When an amphitheater was dedicated in his name in 1918, a speaker summarized Cheever's contributions to the hospital: "He freely gave the best hours of the best days and years of his life to its service."[1]

Cheever was not the only conscientious surgeon on the staff. For example, it was the custom at the city hospital in the 1880s that, whenever a diphtheria patient was threatened with suffocation, an ambulance would be dispatched for one of the senior surgeons. He promptly left his private patients for the hospital to perform a tracheotomy. The medical service was equally fortunate in the quality of its attending physicians. Two of them, Robert T. Edes and George B. Shattuck, were among the thirteen founders of the prestigious Association of American Physicians. Shattuck later became president of the hospital's medical board and for many years served as editor of the *Boston Medical and Surgical Journal*. The careers of most of the medical staff were cast in the same mold.[2]

All thirteen of the original staff were graduates of Harvard Medical School, as were many of the later staff members for years to come. They represented mostly the upper classes of Boston, and many of them moved in the same social circles as the trustees of the Boston City and Massachusetts General Hospitals. But just as the Irish pushed the Boston Brahmins off the board of trustees and the city council, so they began to find their way onto the medical staff of the Boston City Hospital. An example of this new breed was Michael F. Gavin, one of the early house officers. Born in Ireland and educated at Harvard Medical School, he served several years on the staff in a junior position. When he did not advance fast enough, he secured membership on the board of trustees in 1879 and several years later gained appointment to the hospital's surgical staff. Boston's Irish found Tufts Medical School, which opened in 1893, more accessible than Harvard, and eventually many members of the Tufts faculty were on the city hospital's staff, and students from both schools were being taught there.[3]

The medical staff of Cook County Hospital was more uneven in quality, as might be expected in a rapidly expanding city where the atmosphere was filled with youth, vigor, even brashness. The tactics of the politicians were sometimes imitated by the maneuvers of the doctors. For instance, in order to get the Cook County Hospital started after the Civil War, one doctor got himself elected to the county board of supervisors. He succeeded in his objective, but later, when politicians were manipulating the hospital for their own interests, many a staff member ruefully recalled the words of Scripture: "They that take the sword shall perish with the sword."[4] Political maneuvering within the medical staff bore this out. Rush Medical College and the newly formed Chicago Medical College, later the medical school of Northwestern University, were bitter rivals. The original medical staff of Cook County Hospital was composed of three members from each of these schools and seven doctors on neither faculty. When a nonfaculty member died, Edwin Powell resigned his post as a professor at Rush, obtained the vacant position, and promptly rejoined the Rush faculty. This piece of chicanery produced so much discord between the two schools that the county commissioners dismissed the entire board and appointed a new one that gave neither school any representation. According to Quine, Powell's action led to the destruction of the medical board, which had been a self-governing and self-perpetuating body, and Powell was responsible for the "transformation of a noble institution nobly administered into the toy of politicians and the scandal of the medical profession."[5]

When the hospital was rebuilt at its present site, the concession was made that one-third of the attending staff would be nominated by each of the schools and the remaining third filled with nonaffiliated doctors appointed by the county commissioners. In 1881 the politicians began to take matters into their own hands again, appointing homeopaths to one-fifth of the staff positions, and in the following year they interfered with some surgical experiments. The entire staff resigned, and from then on none of the medical schools had any direct representation. Improved selection of the attending staff did not occur until 1905, when the state legislature required that appointments be based on a civil service examination. This action was part of a national movement that introduced civil service examinations in charitable institutions instead of the prevailing method of appointing employees through patronage and assessing these employees for the expenses of politicians in subsequent elections.[6]

Fortunately, Cook County Hospital continued to beckon to some talented doctors. One of the most outstanding was Christian Fenger, who reached Chicago in 1877 after receiving medical training at the University of Copenhagen and further experience in Vienna, Copenhagen, and Egypt. He established a private practice and bought a place on the Cook County staff, where he served as pathologist for two years and then as an attending surgeon.[7] Characteristic of the man were his capacity for long hours of hard work; a craving for new knowledge, whether obtained at the operating table, in a post-mortem room, or from the medical literature; and a drive to apply that knowledge promptly, whether it be a new technique such as antisepsis or asepsis or a complicated surgical operation devised by others or by him. He was devoted to his work at the county hospital despite the demands of his private practice at four other hospitals and his teaching at three medical schools. In these respects he was characteristic of the restless, pioneering, ambitious spirit of Chicago and its doctors. But his personal warmth, his humility, and his genuine desire to help colleagues and house officers at the sacrifice of much time and comfort set him apart from others on the attending staff. Examples of his qualities abound: Fenger teaching at the autopsy table with such brilliance that two of Chicago's budding surgeons happily gave up their lunch hours to assist him; Fenger in his underclothes explaining to house officers the intricacies of the operation he had just performed; Fenger devising new operations on lungs, nervous system, abdomen, and urinary tract; Fenger cursing because his rough technique at an operation on the thyroid gland had temporarily injured a patient's recurrent laryngeal nerve and produced hoarseness; Fenger castigating himself openly for missing a diagnosis; and Fenger forever teaching, teaching, teaching. Small wonder that a generation of Chicago's doctors worshipped him, that over six hundred of them attended a testimonial dinner for him on his sixtieth birthday in 1900, and that a bronze plaque at Cook County Hospital honors him today.[8]

One of the embryo surgeons who had been entranced by Fenger's teaching at the autopsy table was John B. Murphy, who later became an internationally famous surgeon and Chicago's best-known doctor. Colleagues and students admired Murphy for his surgical judgment, which at times seemed uncanny, and his superb skill at the operating table, but they were antagonized by his overweening ambition and burning desire for fame and wealth, which he sometimes achieved by skirting conventions; for example, he gained nationwide publicity when all the victims of the Haymarket Square riot somehow were assigned only to his service at Cook County Hospital. Consequently, even though he was a brilliant surgeon, a popular teacher, and the originator of several operations, few people liked him. James B. Herrick, a professional colleague and himself internationally famous for his clinical descriptions of sickle cell anemia and coronary thrombosis, believed that Murphy failed to attract people personally or to leave any successors because he "used men and knowledge and opportunity as tools and agents for his own ends. He gave not of self but to self."[9] Men like Murphy were often found on the practitioner-dominated staffs of city hospitals in the past. Without strongly organized clinical services to command their loyalty, they tended to build for themselves, not for the institution.

Fenger and Murphy were highly gifted professionally, but they differed profoundly in personal attributes. Their careers point up a characteristic of many large city hospitals: they had room at the same time for doctors of varied abilities and vastly different personalities. So urgent was the need for laborers in the vineyard, so many and so diverse the fields in which they could work, that the hospital could accommodate them all.

The old-line city hospitals that had evolved from almshouses also had their share of superior doctors. An intern at Bellevue in 1875, William Henry Welch, later dean of Johns Hopkins Medical School, wrote, "We have the best physicians and surgeons in the city," and in the following year William S. Halsted, later professor of surgery at Johns Hopkins and teacher of many of America's leading surgeons, selected a surgical internship at Bellevue because of the outstanding men in that discipline. He later recalled that several internists had attracted him as much as the surgeons.[10] One of these was Edward G. Janeway. After interning at Bellevue, he spent six years as curator of the pathological museum, during which he literally "lived in the dead-house," to the dismay of his friends, who thought he was throwing away his chances for a practice. But as an attending physician at Bellevue, beginning in 1872, he built a clinical superstructure upon this foundation of pathology. The two together gave him such uncanny skill in diagnosis that people said he could see inside a patient. He became professor of medicine at Bellevue Hospital Medical College and later at New York University, where he was also dean, and the city's leading consultant for over two decades.[11] His son, Theodore C. Janeway, was, in his turn, an outstanding professor of medicine, who later headed the department at Johns Hopkins. When he became attending physician at New York's City Hospital on Blackwell's Island in the first decade of the twentieth century, the service had been neglected, "the internes were poor...and hard to obtain. In a short time, everything was changed." Under his leadership it became an active, effective service, and candidates eagerly competed for internships.[12] Father and son epitomized the successful teachers of that era when a medical or surgical service was built around a single man.

By the mid-nineteenth century serving at the Philadelphia Hospital had become a tradition among Philadelphia's doctors. Its rolls were filled with famous names, including Gross and Agnew, well-known surgeons of Jefferson Medical College and the University of Pennsylvania (see Chapter 1). Among its outstanding physicians was William Osler, whose brief stay in Philadelphia illustrated well the possibilities of a city hospital. Arriving in 1884 as professor of clinical medicine at the University of Pennsylvania and finding that the clinical material at the university hospital "was limited and so was the spirit," he concentrated his work at the Philadelphia Hospital, where he found fertile and almost limitless pastures. Here he added to the store of knowledge that later filled the many editions of his notable textbook of medicine. Here he fed the flock of students that trailed him in wards and autopsy room, teaching them to "observe, record, tabulate, communicate," and demonstrating by his example that "medicine is learned by the bedside and not in the classroom." Here his experience with patients, students, and coworkers nourished the spiritual flame that made his later talks and writings the inspiration of doctors worldwide. He left Philadelphia in 1889 to become the first professor of medicine at Johns Hopkins Medical School.[13]

Although city hospitals in the nineteenth century attracted bright, energetic, talented doctors to their staffs, we cannot assume that all or even a majority of the medical staffs of these hospitals were of that stature. Very little is written about the others, but they were there. Some were political hacks, who were neither conscientious nor knowledgeable; others were diligent workers of limited ability, who were content to employ their single talent well and be counted as colleagues of the stellar performers. Sometimes, however, these doctors attempted tasks beyond their competence. Occasionally the results were ludicrous, as in a story told by a former intern at Cook County Hospital: Charles Gilman Smith owed his staff position to his engaging personality and his friendship with leaders of the staff, but his busy private practice left him little time for the city hospital or for keeping up his surgical skills. One day when he was clumsily trying to remove a hemorrhoid, the patient kicked him in the face and broke his glasses. After that he relinquished all pretense of active duty on the staff.[14]

Unfortunately, when others ventured more than they were trained for, the patients did not always escape so readily; nor did the bunglers always take themselves off the staff. Colleagues and interns had to prevent what damage they could, repair what could not be prevented, and cover up what could not be repaired. Interns also tried to avert catastrophe by asking an abler doctor to help with a diagnosis or maneuvering matters so that difficult operations could be performed by competent surgeons. For example, interns at Charity Hospital in New York City used to conceal the need for certain operations from the patients' attending physicians. Then, when Halsted arrived, they would label the case an emergency so that he could perform the operation. But such tactics could not right the wrongs of a whole hospital, and many a patient must have been victimized by blind ambition or gross stupidity.[15] Nor were incompetence or lack of interest the only faults of attending doctors. A more frequent problem was lack of time. Halsted could not recall learning anything from two of the three surgeons on his service while an intern at Bellevue in 1876, for "they were irregular in attendance and entrusted almost everything to the interns."[16]

The system by which the best practitioners were encouraged to devote many hours to patient care and teaching in city hospitals also required them to make their living elsewhere in private practice. But the best doctors were the most sought after by private patients and also the ones who desired time for study. Conscientious doctors tried to cope with the overload in several ways: by starting work at dawn and keeping it up until midnight, by employing assistants for some of their private work, or even by avoiding as much as possible the private practice that brought in the "damn guinea," and thereby flattening their pocketbooks. Osler used the latter method in Philadelphia. Practically all the doctors on the faculties of the two leading medical schools, Pennsylvania and Jefferson, were in the private practice of medicine, including those teaching the basic sciences. Instead, Osler wanted to spend all the time he could learning and teaching in the hospital. A colleague later wrote of him: "First, then, we ... sought to make a practitioner of him. But of that he would have none. Teacher, clinician, consultant, yes, gladly; but practitioner—no! And that with emphasis."[17] But Osler was a bachelor at the time; others could not sustain themselves on a trickle of income. Instead, most good teachers and would-be teachers were condemned by the system to carry a full burden on each shoulder.

Attending physicians varied also in their concern for patients and coworkers. Some were considerate and kind to all; others were often guilty of tactless, coarse, callous, even cruel behavior toward colleagues, students, subordinates, and especially patients. Unfortunately, such conduct was more likely to be tolerated in city hospitals than in private institutions. But whatever the deficiencies of staff doctors, and however much they were pressed for time, they were seldom removed from office. Doctors prefer to look the other way rather than criticize an erring brother; on the staffs of the city hospitals they tended to ignore the harm he did or do his work for him, even though they grumbled as they did it. They knew, moreover, that there were only so many competent doctors who were willing to neglect their practices for the labor of attending at the city hospital and that if a questionable physician or surgeon were dropped, a worse one might take his place.

The task of the attending physicians would have been impossible had it not been for another group of doctors, the house officers. At first, house officers in almshouses and early hospitals had been medical students, supernumeraries who, in order to pick up a few crumbs of practical knowledge, were willing to perform the task at hand—from nursing a very sick patient all night, compounding pills and mixing potions, to writing progress notes and dressing wounds— tasks that today would be apportioned among nurses, nursing aides, pharmacists, and interns. As internship training in city hospitals became more attractive, house officers were required to be graduates of medical schools. Gradually a definite and fairly rigid system developed. At Cook County in 1889 the house staff consisted of twelve interns. Each served as a junior physician for six months, performing the simpler work of a doctor: writing histories, dressing wounds, and prescribing minor medications. For another half year he served on the obstetrical and gynecological service and as admitting physician. During the third six months he was in charge of his service and responsible only to the attending doctor.[18]

At Bellevue internships were even more specialized. As early as 1856 a prospective house officer selected either a medical or a surgical service, where he spent the entire eighteen months. The experience gained was tremendous and the responsibilities, particularly after the first year, awesome. For all his hard work and long hours the house officer at Bellevue received no compensation for the first year. He even had to live outside the hospital at his own expense. During the last six months he was given his board and lodging but no salary.[19] Other city hospitals were slightly more generous, giving board and lodging to all house officers, but not until the Second World War did all city hospitals begin paying a salary to first-year house officers.

Yet, these positions were eagerly sought. At first internships were sometimes obtained through personal friendships or political influence. For instance, until 1883 each house officer at Philadelphia Hospital was selected by a politician. Later many leading hospitals awarded these positions on the basis of an examination, which was likely to be comprehensive and difficult. In 1905 a civil service examination became a requirement for the house staff of Cook County Hospital and the attending staff as well. A candidate's rank in the examination usually determined not only whether he would be appointed but also whom he would work under. Thus, aspirants for internships strove to place high on the list. In Chicago, either from misguided local pride or excessive chauvinism, only graduates of the city's leading medical schools were eligible to take the examinations, and the candidate who "wrote first in the county exam" was highly honored. At the fraternities of Rush Medical College, for instance, it was the custom for all to rise when someone who had won this position entered the dining room. There was good reason for the tribute, for many who "wrote first" subsequently attained prominence in their profession. In the 1880s these included Murphy, Herrick, the pathologist Ludwig Hektoen, and Frank Billings, who later became dean of the University of Chicago Medical School.[20] Other city hospitals could point to equally famous names among their former house officers.

As city hospitals became larger and techniques of diagnosis and treatment more complex, house officers came to occupy key positions. Theoretically taking all their orders from attending doctors, they were on their own much of the time. From the beginning, some city hospitals arranged for supervision of house officers in the absence of attending doctors by requiring them to report problem cases to a full-time staff doctor. At midcentury the superintendent fulfilled this function along with his administrative duties. Later, when administration took all his time and his medical skills became rusty, or when the superintendent was a layman, house officers went their way much of the time, unchecked and unaided. The editor of the *Medical Standard* wrote of Cook County Hospital in 1901: "There is no executive head, competent to understand and supervise the medical and surgical side of a great charitable institution like this. The institution seems to have been running itself, so far as the interests of the patients was [*sic*] concerned, or left to the control of young internes just out of college." In 1914 Cook County Hospital established the post of medical superintendent to provide medical supervision (see Chapter 5). Other hospitals took similar action, using various titles.[21] But at Cook County and elsewhere these officials often had just finished their own hospital training or were so burdened with administrative chores that they could not keep their medical skills sharpened. Since knowledge was the coin in which the house officer was paid, he wasted little time on doctors whose titles exceeded their competence.

Thus, dual jurisdictions were perpetuated: administrative officers on the one side and the medical hierarchy on the other. House officers tended to believe that administrators were not really interested in making people well but only in getting involved in politics, pinching pennies, and slavishly following a rigid set of rules. Administrators, on the other hand, felt that house officers, in their zeal to learn and practice, were willing to flout every rule to the point of destroying the hospital. These differences caused friction less often than one might expect, probably because each group left the other alone as far as possible, but every once in a while the sparks would fly. One notable example was the firing of Leonard Wood from his internship at the Boston City Hospital in 1884 by Superintendent George H. M. Rowe, who was described by Wood's biographer as "one of the best hospital-managers in the country: honest, devoted to his institution, a master of detail, and fundamentally kind; but fussy, irascible and stubborn." House officers were forbidden to operate on patients except in an emergency and then only with the permission of the superintendent. When Wood, who had been guilty of other infractions of the rules, proceeded to operate without permission, Rowe told him he was through. Whether he would have done so if he had known that the hospital's bad boy would become General Wood and a national hero is uncertain, but probably so, for two decades later Rowe was still complaining of house officers who degenerated into the free and easy manner of the uncouth medical students in Dickens' *Pickwick Papers*.[22]

The real villain of the episode was neither Rowe nor Wood but the system, which used interns to do the work of doctors cheaply but gave them practically no supervision, because competent teachers would have been expensive. If an experienced surgeon who commanded Wood's respect had been supervising the intern at all times, he could have performed suitable operations with proper guidance. As the system actually functioned, however, it attracted house officers like the Cook County Hospital intern who informed Jane Addams that he preferred doctors on his ward who visited irregularly because they did not "interfere with the intern's treatment."[23] Many city hospitals have boasted that they allowed interns to assume extensive responsibilities without admitting the consequences: too much responsibility too soon means too little supervision, which increases the chance of harm to patients.

The house officer's status was ambiguous. Trusted by attending physicians, obeyed by nurses, and often adored by patients, he made decisions that spelled life or death. At the same time, although a medical graduate he was still a student, learning from both doctors and nurses, and his position was temporary: his tenure usually lasted only a year or two. Consequently, he was not consulted about hospital policy, and when he ventured to say something about administration, he was ignored. Occasionally the ambivalent status of a house staff fomented a rebellion against the hospital superintendent. Not until the 1920s or later, when well-staffed university services were operating in many city hospitals (see Chapter 6), did house officers feel that they had someone in authority who understood their problems.[24]

House officers spent long, laborious days and many sleepless nights trying to keep up with the tasks required by a busy city hospital. But they could expect little direct assistance from attending physicians, who were squeezing their few hours at the city hospital out of a busy private practice. The attending staff wanted to retain its status as a select group of doctors who had reached a pinnacle of prestige. Thus, they were loath to share their positions by increasing the number of appointees, even if additional competent doctors were available. But a national movement was developing that indirectly forced attending doctors to divide up their responsibilities.

The last half of the nineteenth century saw the development of medical specialization in the United States. Two conditions necessary for specialization were fulfilled by then: the aggregation of knowledge and the development of special instruments in individual fields, and the growth of cities to the point where the population could support a doctor who confined his practice to one category. These in turn produced a third necessary condition: the development of institutions to facilitate the practice of specialties, especially hospitals for particular types of illness and specialty societies.

Since most specialty hospitals were private and highly dependent upon patients' fees, they could not care for the bulk of the sick poor. Thus, pressure was placed on city hospitals to develop specialty services. National societies for promulgating knowledge in a special field and promoting its interests developed early. Table 4 shows how their founding coincided closely with the establishment of special wards for particular diseases at one city hospital.

**Table 4. Beginnings of specialty societies in the United States and specialty services at the Boston City Hospital.**

| Society | Year founded | Year service started at Boston City Hospital |
|---|---|---|
| American Ophthalmological Society | 1864 | 1864 |
| American Otological Society | 1868 | 1869[a] |
| American Neurological Association | 1875 | 1877–1878 |
| American Dermatological Association | 1876 | 1868 |
| American Gynecological Society | 1876 | 1892[b] |
| American Laryngological Association | 1879 | 1869[a] |
| American Surgical Association | 1880 | — |
| Association of American Physicians | 1885 | — |
| American Orthopedic Association | 1887 | 1928[c] |
| American Pediatric Society | 1888 | 1923[d] |
| American Urological Association | 1902 | 1958 |

Sources: William G. Rothstein, *American Physicians in the Nineteenth Century: From Sects to Science* (Baltimore: Johns Hopkins University Press, 1972), p. 213; David W. Cheever et al., *A History of the Boston City Hospital from Its Founding until 1904* (Boston: Municipal Printing Office, 1906), pp. 154–162; John J. Byrne, ed., *A History of the Boston City Hospital, 1905–1964* (Boston, 1964), pp. 151, 227, 232–255.

a. The ear clinic gradually became the clinic for diseases of the ear, nose, and throat.

b. Outpatient department, 1973.

c. Consultant, 1903.

d. Special wards had been set aside for children in 1919.

Just as the hospital at the Philadelphia almshouse had been separated into medical and surgical wards by 1811, so the wards of the Boston City Hospital had been divided in the same way at its opening in 1864.[25] Patients were culled from those wards as each new specialty was recognized, the remaining patients eventually making up the general medical and surgical services. Because general surgeons and internists recognized themselves as specialists later than representatives of the narrower specialties, their national societies were slow in forming, the first being the American Surgical Association in 1880 and the Association of American Physicians in 1885.

Ophthalmology, in contrast, was perceived as a specialty in the 1860s and was made a separate service when the Boston City Hospital opened. Gynecologists at that hospital tried to form their own service in 1873 but were allowed only to open a clinic in the outpatient department. Probably this was because many general surgeons considered abdominal operations on women to be in their province. Likewise, general surgeons also vied with orthopedic surgeons in treating fractures. Only a consultant in orthopedics was permitted at Boston City at first, and even that position was not established until 1903, although Bellevue began an orthopedics department in 1875 and the Philadelphia Hospital in 1899.[26]

Patients with venereal diseases were isolated from others because some were contagious and all bore a social stigma. Out of these wards grew genitourinary services, though often much later than other specialty services. For a long time many of the most complex operations in this field were performed by general surgeons. Fenger, for instance, devised several important urological procedures. Preemption of the field by general surgeons probably explained why the department of urology was not organized at Cook County Hospital until the 1920s and not at the Boston City Hospital until 1958.[27]

Internists also clung to some of their prerogatives, including pediatric medicine. The American Pediatric society was founded in 1888, and Cook County Hospital appointed its first staff of attending physicians for a children's ward in 1890, while children at most city hospitals were placed on the same medical and surgical wards as adults for some time after that. A pediatrics service was not officially recognized at the city hospitals of Boston and Baltimore until 1923 and 1935, respectively.[28]

Other city hospitals responded less promptly than the Boston City Hospital to the pressures of the specialists. At Bellevue an attempt to establish special services in dermatology, neurology, otology, ophthalmology, and orthopedics was beaten back in the medical board in 1863. In 1875 the board yielded to the extent of forming departments of orthopedics and gynecology. The Philadelphia Hospital established its first specialized departments in dermatology, neurology, and ophthalmology in 1877.[29]

At Cook County Hospital in the 1860s meetings of the medical staff were filled with "impassioned harangues for and against recognition of the specialties." The staff was loath to split up the general medical and surgical services and for a long time used the device of appointing consultants in the specialties, beginning in 1866 in dermatology, gynecology, and diseases of the eye and ear. Not until 1915 was a special ward opened for diseases of the ear, nose, and throat, and in 1916 the first residency in ophthalmology was started. In fact, as late as 1932 one of the major criticisms in a survey of that hospital was the need for more specialization.[30]

In addition to new knowledge, specialty services brought to city hospitals a fresh, vigorous group of doctors, whose enthusiasm for new, rapidly developing fields of medicine was contagious. The specialists threw their energies into their city hospital services all the harder because the medical profession outside the hospital was strongly resisting their progress. General practitioners believed that specialists competed unfairly with them because they claimed to offer services not provided by most doctors and because some of them advertised to the public.[31]

One specialty received particular attention in city hospitals. As cities grew larger and more crowded, contagious diseases increased, and as more people lived in crowded quarters, more victims of these diseases needed hospitalization. From the beginning local governments had been forced to provide facilities during epidemics, usually frame shacks hastily erected or abandonded buildings summarily commandeered. Some of these were kept on as more permanent hospitals. Smallpox patients, especially, were placed in quarters such as these. As late as 1872 the Board of Health of the District of Columbia described the "shanties of rough boards . . . not even weather tight" that had been added to the original smallpox hospital, itself only two rooms of unfinished boards, and commented that "a man must indeed be alone and forgotten in this world who is willing to enter such premises."[32] The present site of Bellevue was first a smallpox hospital to which the other parts of the almshouse-hospital complex were added. Likewise, in the nation's capital the other components of the hospital—the almshouse and the city jail—were built adjacent to the original smallpox hospital. In Boston a building for smallpox patients was erected on the city hospital grounds in 1865, while Chicago maintained a separate hospital for smallpox up to the 1950s or later, even though no cases had been diagnosed for years.[33]

The almshouse and its successor, the city hospital, were also forced to take patients with other, less dreaded contagious diseases, such as scarlet fever, diphtheria, and measles, because private hospitals excluded them. At first these patients were banished to an outlying ward in makeshift quarters, where it was difficult or impossible to isolate those with various communicable diseases from each other. Eventually, specially designed buildings were erected for contagious diseases. At the Boston City Hospital temporary wooden buildings erected adjacent to the city hospital grounds in 1892 were transferred to city hospital control in the following year, and in 1895 a new building, named the South Department, was opened for the treatment of contagious diseases; there nearly twenty-five thousand patients were treated in the next ten years. As a result of the pioneer use of diphtheria antitoxin and the expertise of John H. McCollum and his staff, fatality rates from diphtheria fell from 46 percent prior to 1895 to 12 percent in the next five years.[34]

Such decisive results in diphtheria, even though they were not duplicated for other diseases until a half century later, gave a strong impetus to the establishment and maintenance of first-rate facilities for contagious diseases in American cities. Contagious disease units were placed within the city hospital in a few cities, such as New York and Chicago, although these did not reach the high standard set by the Boston City. Rather than upgrade the meager facilities for contagious diseases in city hospitals, many communities chose to follow the example set by Glasgow in 1865 by establishing a separate contagious disease hospital. Both the health authorities and the public urged this course. Health departments wanted to run a hospital of their own, which seemed logical because these departments made and enforced the rules throughout the city for the control of communicable diseases. The public, reasoning that germs that could travel twenty feet could also spread for miles, wanted these hospitals banished beyond the city limits. Furthermore, all would benefit if a separate institution, without the stigma of almshouse or city hospital, were available to upper- and middle-income patients. Thus, New York City established Riverside Hospital for this purpose in 1875 and Willard Parker Hospital in 1885. Philadelphia had provided a separate hospital for infectious diseases as early as 1810, and this became a permanent institution, the Municipal Hospital for Contagious Diseases, in 1865. Baltimore built a separate hospital for communicable diseases in 1909 during a wave of reform in politics and crusading in public health. Washington, consistent with its custom of giving funds to private hospitals to perform the city's tasks, in 1900 erected buildings at Providence and Garfield Hospitals for patients with contagious diseases and contracted to pay for care given there.[35]

It is interesting to consider what would have happened if all city hospitals had established their own units for contagious diseases from the beginning. Proper isolation procedures would have saved lives, for the makeshift system of merely designating certain wards as fever wards without establishing rigid isolation procedures allowed infections to spread to other wards.[36] Furthermore, where a contagious disease unit was established as part of a general hospital, doctors and nurses gained experience in the management of patients with contagious diseases and in the rules of strict isolation. Instead of these diseases being out of mind because out of sight, this branch of knowledge would have been better taught, and the increased attention would probably have stimulated more research, as happened with typhoid, typhus, and pneumonia. Patients with these infections had been admitted to the wards of city hospitals all along.

Another special department, pathology, was usually one of the strongest in city hospitals, especially because so many patients suffered from unusual or advanced conditions less often seen in private hospitals. At first, post-mortem examinations were performed in American hospitals by clinicians. Years at the autopsy table produced knowledge and sharpened the skills needed for the diagnosis of disease in the living; most of the outstanding diagnosticians of the period had spent considerable time there, for example Edward Janeway at Bellevue and Fenger at Cook County. Likewise, Osler spent endless hours, including all day on Sunday, at the autopsy table at the Philadelphia Hospital.[37]

Pathology was valuable not only for training the individual doctor but, of more importance, for providing an understanding of the genesis, behavior, and effects of diseases; in other words, it was the basis for all of medicine and surgery. Thus, it deserved its own full-time disciples. Toward the end of the nineteenth century a few doctors began to make pathology a career, and several of them were involved directly or indirectly with city hospitals. William Welch, after two years' training in pathology in German universities, opened a pathological laboratory at the Bellevue Hospital Medical College in 1878 and obtained material for his studies and teaching from the adjacent Bellevue Hospital. This was the second laboratory of its kind opened in the country, the first having been established the year before by T. Mitchell Prudden of the College of Physicians and Surgeons in New York. Most of his teaching material was probably also obtained from Bellevue. Because of the reputation Welch developed at Bellevue, he was called to head the department of pathology at Johns Hopkins in 1884. His first assistant was William T. Councilman, who was pathologist at Bay View Asylum. After performing an autopsy at the city's almshouse-hospital, Councilman would transport his specimens to the laboratories of the Hopkins department of pathology on a tricycle; Welch later recalled that his assistant was nearly arrested once because the tricycle fell over and spilled all the specimens out onto the street. According to a colleague, Councilman was the first in the United States to confirm the discovery of the malarial parasite made by Alphonse Laveran, a French army surgeon; the evidence probably came from patients at Bay View Asylum.[38]

When Councilman left Baltimore for Boston, he spread the influence of the Hopkins pathology department to Harvard and the Boston City Hospital, while three of his successors at Bay View carried it elsewhere: Eugene Opie became professor of pathology at Washington University in St. Louis, George H. Whipple at the University of California and later Rochester, and Milton C. Winternitz at Yale. Whipple and Winternitz eventually became deans of their medical schools, and Whipple also won a Nobel Prize.[39] It seemed for a while that Bay View was on the road to rivaling the celebrated Allgemeines Krankenhaus in Vienna as a birthplace of pathologists.

Until Councilman's arrival the heads of the Harvard department of pathology had used this science as a stepping stone to a consulting practice in internal medicine. In 1892 Reginald H. Fitz, pathologist at the Massachusetts General Hospital, gave up the chair in pathology at Harvard when he was appointed to a similar one in medicine, and Councilman succeeded him. Councilman shifted the base of operations to the Boston City Hospital, where he organized a laboratory that became world famous. Its distinguished reputation was carried on under his student and successor, Frank B. Mallory, who trained 123 pathologists, twenty-three of whom became professors of pathology in medical schools.[40]

While the special branches of medicine were developing in city hospitals, the medical sects also sought admission. The most influential of these was homeopathy, which had been introduced in the first half of the nineteenth century and was highly popular in the United States at midcentury and later. In a private hospital the influence of the regular profession upon the board of trustees was usually sufficient to keep out the homeopaths, who were forced to open their own hospitals. But this sect clamored to get on the staffs of city hospitals, arguing that public hospitals were for the public and that the patients should have access to homeopathic treatment if they wanted it. As long as the medical profession could not conclusively demonstrate the superior therapeutic value of regular medicine over homeopathy, laymen had difficulty choosing between the regulars and the homeopaths. Consequently, the decision as to who should staff the city hospital became a matter of influence, and to politicians homeopaths were a constituency with votes.

Pressure on politicians was particularly strong in Chicago; Quine speculated that in the 1850s the city contained two-fifths of all the homeopaths in the world. Their influence had precipitated a dispute in 1857 over the staffing of the city's new hospital for the sick poor, causing the building to remain vacant for two years (see Chapter 2). The medical staff of the original Cook County Hospital was drawn from the regular profession, but in 1882, after staff appointments had come under political control, homeopaths were given charge of one-fifth of the beds. Later, beds were also allocated to physicians of another sect, the eclectics, who claimed to use the best drugs from all systems of medicine but actually relied on botanical remedies. In 1898 there were fourteen regular, five homeopathic, and five eclectic physicians on the medical staff. Despite many protests, the medical profession did not dislodge the sects from the staff until appointments were placed under the civil service system in 1905.[41]

In the eastern states the regular profession remained firmly entrenched in city hospitals despite repeated assaults by homeopaths. When the sect sought admission to the staff of the Boston City Hospital in 1886, the board of trustees answered that the hospital had been successful with the physicians it had, that the introduction of a conflicting method would cause trouble, and that special wards would produce administrative difficulties. How much of the board's decision was the product of careful reasoning and how much resulted from close social relationships between the board of trustees and their fellow Brahmins on the medical staff can only be surmised. The faculty of Boston's homeopathic medical school, even though it was connected with Boston University, continued to be excluded from the Boston City Hospital staff until after the school was accepted as a regular medical school.[42]

The Philadelphia homeopaths made their bid early. In August 1859 the faculty of the homeopathic college offered to take care of the medical department at the Philadelphia almshouse and to furnish drugs free. It happened that the practice of electing a board of guardians had just been discontinued. The new twelve-member board appointed by the courts contained, at least in the opinion of the regular profession, "men of enlarged liberal views, conjoined with superior practicable [sic] ability," and it had no difficulty in rejecting the offer of the homeopaths.[43] As the influence of homeopathy waned with the waning of the century, pressure for representation on the staffs of city hospitals lessened, and it stopped altogether in the early twentieth century, when homeopathic schools either closed or conformed to the curricula of the regular schools.

In the nineteenth century women doctors fared poorly in city hospitals as in most others in the United States. Although Sarah Adamson served as an intern at the Philadelphia Hospital in 1851, her responsibilities were confined to obstetrics and gynecology. The first city hospital to appoint a woman to the attending staff was Cook County in 1881, and the same year that hospital appointed its first woman intern. The following year Philadelphia's city hospital appointed a woman attending physician and in 1883 a woman intern; three graduates of the Woman's Medical College of Baltimore were appointed to Bay View Asylum in 1891. But these hospitals did not consistently appoint women from that time on; for instance, a second woman was not appointed to the staff of Cook County Hospital until 1899. Other city hospitals were more recalcitrant: in 1901 only six women held positions on the attending staffs of city hospitals, compared with 136 on the staffs of private general, maternity, and children's hospitals in the United States. Corresponding figures for house officers were four and forty-two. The shortage of house officers during the First World War opened the door to women interns in some hospitals, but the Boston City Hospital—as conservative here as it had been progressive elsewhere—did not appoint a woman house officer until 1931. Yet, it was still in the minority of all hospitals, since in the mid-thirties only 105 of 712 American hospitals approved for internships accepted applications from women. The best we can say about the attitudes of city hospitals toward women doctors is that they were no more old-fashioned than most general hospitals.[44]

In the half century following 1860, while city hospitals were trying to keep pace with the growth of cities and the increasing complexities of medical care, methods of medical education in the United States were also changing profoundly. City hospitals participated intimately in these changes. They attracted medical schools like magnets. At one time there were five clustered around Cook County Hospital and three around Bellevue, and in 1883 the expanded Harvard Medical School was placed equidistant from the Boston City Hospital and the Massachusetts General Hospital so as to maintain good relations with both. After Rush Medical College and Cook County Hospital were destroyed in the Chicago fire, the college waited four years until the location of the new county hospital was settled and then built its school across the street. Nor did medical schools hesitate to use the city hospitals for purposes of publicity. In 1895 diplomas of Rush Medical College carried a picture of Cook County Hospital and certified that the medical graduate had attended Rush clinics at that hospital.[45]

In the decades after the Civil War aspiring doctors flocked to schools connected with public hospitals in large cities, attracted by the number and variety of patients to be found there. Walter Reed, after receiving his M.D. from the University of Virginia in 1869, attended Bellevue Hospital Medical College, which had opened in 1861, to obtain needed clinical training.[46]

Yet, relations between medical schools and city hospitals were tenuous. At midcentury the schools had access to the hospitals on sufferance, as in most private hospitals. At the period's end a few city hospitals had ventured a step or two with medical schools, but the partners seldom danced for long to the same tune. Exceptions included Bellevue, where each of three medical schools had some supervision over one of the four medical and surgical divisions, and Boston City Hospital, where Tufts and Harvard had each been assigned one medical service. A close relationship existed between Charity Hospital in New Orleans and Tulane University, although full control was not vested in the school except in one gynecological ward. As late as 1910 Abraham Flexner concluded that only the city authorities in Toronto had granted to a medical school (the University of Toronto) complete medical control over a sufficient number of patients (about five hundred) to satisfy educational requirements, and he held this up as a model city-university arrangement. In contrast, most medical schools had little voice in the conduct of patient care in city hospitals, and many students' only glimpse of a ward patient was in a demonstration clinic once or twice a week.[47]

One reason for these conditions was that the medical schools seemed to expect little more. They were proprietary institutions, in spirit if not in fact, and their main interest was to get their faculties onto the hospital staffs or the hospital staffs onto their faculties; the teaching itself they were content to leave to the individual doctor. If they needed any barometer of the students' preferences, it was provided by the number who registered for each course; and since standards were low, amphitheater walls were elastic, and students' fees were divided up among the teachers, the schools accepted just about all who applied.

The number of medical students in the United States increased progressively each decade from approximately five thousand in 1860 to approximately twenty-five thousand in 1900, or more than twice as fast as the population.[48] The medical schools packed them into amphitheaters, where students established their own rules to assure that seniors occupied the front seats.[49] In 1848 Charity Hospital in New Orleans completed an amphitheater seating six hundred. Although an English surgeon claimed to have seen "at least a thousand" students attending a lecture in Bellevue's amphitheater in 1874, it was replaced with a larger model in 1890. A large amphitheater was included in the hospital opened by Cook County in 1866, and it is typical of medical education of the era that Rush Medical College was able to carry on without hospital beds for several years after the Chicago fire by using the amphitheater of the former county hospital for lectures.[50]

At first hospital amphitheaters were used mostly by surgeons.[51] The rapid surgery necessary in the days before anesthesia encouraged the spectacular, and the custom remained after general anesthetics were introduced in the 1840s. Accomplished surgeon-teachers could make the words flow as their fingers flew, and larger audiences merely stimulated greater showmanship. Later, professors of medicine also began to "show and tell" before large audiences.

The popularity of the large clinic grew progressively. At the College of Physicians and Surgeons in New York City college-wide clinics began in 1841. They were increased to three per week in 1856 and to ten per week in 1866, and in 1869 clinical professors were appointed to take charge of them.[52] The big clinic at its best was able to impress many facts and principles on the minds of medical students. The amphitheater clinics of Jacob M. DaCosta, which began at Jefferson Medical College in 1866, were considered "models of the finest methods of medical education—clear, systematic and impressive." At Cook County Hospital the big clinic was especially glorified. As late as 1903 Herrick, one of Chicago's best teachers, was still vigorously defending teaching one or two hundred students in an arena clinic. Yet even he admitted that when the students' attention lagged, the teacher often yielded to the temptation to entertain his audience, to be "amphitheatric," to "burn red fire."[53]

Ideally the big clinic called for a master physician or surgeon to demonstrate and explain the features of the disease presented by the patient exhibited in the pit, but if the case and description did not match, no one need be the wiser. William Pepper at the University of Pennsylvania in the 1880s was known "on occasion to give a brilliant discourse to the students on Addison's disease, using a patient with ordinary jaundice for the purpose of the clinic, knowing full well it was a deception."[54] Students could not observe and examine patients for themselves. Occasionally a student was called to the pit to examine a patient, but David Riesman, who graduated from a leading school, the University of Pennsylvania, in the 1880s, could not recall examining more than two or three patients during his entire medical school career. An 1891 graduate of the College of Physicians and Surgeons of New York wrote ruefully, "We never came within a mile of touching a patient."[55] Sometimes a student in the pit was quizzed by the professor. He strained and sweated while his fellows clapped or jeered. The professor mystified, then clarified. But though he searched out the path to the correct diagnosis amidst the fog of the students' ignorance, it was *his* knowledge that found the path. Thus, his pronouncements often failed to help his listeners in the management of their patients' illnesses when they were on their own. Graceful generalizations could not replace close contact with patients.

The irony was especially galling when amphitheater students were passive spectators in city hospitals while hundreds of patients lay on the wards in need of someone to pay attention to them. By discarding the preceptorship and adopting instead the passive learning of the large clinic, the professors ignored the advantages of learning by doing. They forgot that in an applied science, such as medicine, knowledge flows not only from science to practice but from practice to science as well. It is significant that this emphasis on mass education was accompanied by very little clinical investigation; only when medical education became individualized did clinical investigation flower in the United States.

In some hospitals large clinics were supplemented by ward walks, which were supposed to bring students into close contact with patients. No one perceived their limited value better than Mike Flaherty, a patient in Cook County Hospital around the turn of the century. After an enthusiastic young attending physician and his large retinue had bustled through the ward, Mike called to the man in the next bed, "Ole, we ought to be a hell of a lot better. The professor has just walked by."[56] Ward walks could be exercises in futility as students strained to hear, craned their necks to see, and shifted from one aching foot to the other as the teacher looked, listened, examined, and explained. The key to optimal learning was absent: the students were not closely involved in the care of the patients.

In fact, at the end of the nineteenth century, in some hospitals, students were not allowed in the wards at all. A survey conducted by the Boston City Hospital in 1886 showed that 19 percent of twenty-seven public hospitals and 33 percent of ninety-one private hospitals did not permit students on their wards. Some conscientious students, who knew better than their teachers what they needed, obtained a post as house officer to gain clinical experience. At Harvard in 1888 eight hospitals, including the city hospital, were approved for residencies for the few senior students fortunate enough to get them. As late as 1910 patients were demonstrated to students at Cook County Hospital only after they were wheeled into a side room. Elsewhere some attempts were made to bring students into contact with patients. At the University of Pennsylvania in 1877, besides twenty-four hours of amphitheater clinics, students attended five hours of small specialty clinics and eight hours of instruction on the wards in small groups. Much of early small-group teaching occurred in the specialties for two reasons: since the specialties were considered of minor importance, only a few students at a time were assigned to those subjects; and since each specialty focused on one or two diagnostic or therapeutic instruments, instruction in their use required individual or small-group teaching.[57]

From the lecture and the big clinic the chief method of teaching clinical medicine eventually shifted to the clinical clerkship. In this system the third- or fourth-year medical student was assigned to a team of attending and resident physicians and interns. He was taught by this team and under their supervision took certain responsibilities for medical care of patients. An early stage in the evolution of clerkships began at Charity Hospital of New Orleans. In 1848, when the wards of this hospital were assigned to the Medical College of Louisiana (later Tulane University) for part of the year, the college did not initiate clinical clerkships, although it did allow students to accompany professors on their rounds. In 1856, however, when certain wards were allocated to the rival New Orleans School of Medicine, its students were given free access to the wards and instruction at bedside. Yet even these students were not really clinical clerks, because they were not members of the medical care teams. However, before this school could start a trend in medical education, the Civil War destroyed it.[58]

In the United States genuine clinical clerkships began at Johns Hopkins Hospital in 1896. As a new school with its own hospital, Hopkins could make its own traditions; others were hampered by the encrustations of the hospitals they used for teaching. The attitude at Harvard at that time was expressed openly: "It will never be possible, nor it is desirable, to have students turned loose in a ward in contact with patients." Immediately after the Civil War teaching of students by the Department of Medicine at Harvard had consisted of two lectures each week, one conference, and, for contact with patients, six ward walks with an instructor while he was making rounds. Later even this contact was cut off, when students were excluded from ward walks at the Massachusetts General Hospital from 1897 to 1906 and at the Boston City Hospital from 1902 to 1906. Work with patients in the outpatient clinic was intensified to make up for this lack of experience. Sometime around 1900 students were taken by instructors onto the wards of the Massachusetts General and Boston City Hospitals and taught specifically how to examine patients and treat their illnesses, and in 1906 groups of students were allowed to examine a patient in a lecture room while the professor was lecturing about the case. Around the same time one of the young Turks, Henry Christian, developed an elective clinical clerkship for Harvard students at the Massachusetts General and Carney Hospitals, but it was not until 1914 that regular clinical clerkships for all students were established at the Harvard teaching hospitals, including City Hospital. Correspondingly slow progress was made in other medical schools, some of which did not develop clinical clerkships until the late twenties.[59]

It would be gratifying to report that city hospitals pioneered in the reforms that led to clinical clerkships, but this was not so. Too many were like the St. Louis City Hospital, which Flexner said "has a medical and surgical staff who do no teaching and a teaching staff who do no doctoring."[60] Real clinical clerkships could not exist until each service functioned as a unit, from the professor at the top to the clinical clerks at the bottom. The secret of effective clerkships was that students were not simply turned loose on a ward; rather, everyone above them recognized his responsibility to supervise as well as to teach them. Thus, university-type services had to be developed before proper clerkships could be established, and such services did not come in many city hospitals until well into the twentieth century (see Chapter 6).

In sum, profound changes occurred in city hospitals after 1860. Up to this time medical staffs were composed of doctors of varying ability, working in the hospital—as they did in private practice—as individual practitioners. By 1910 staffs had become loose confederations of doctors with some common interests and aims but no central direction or control of medically related affairs; they were affiliations rather than unions. Loyalties still tended to cling to individuals, especially the stellar performers, rather than groups. The practitioner-doctors advanced medical care in the city hospitals by absorbing the rapidly moving advances in medical science, by applying them in the hospital according to their individual ambitions and abilities, and by teaching colleagues and house staff on an informal basis.[61]

During this half century interns progressed from ad hoc assistants to graduate doctors who were assigned responsibilities step by step according to their experience. The education of medical students in city hospitals changed from a loose, observer-apprentice system to teaching in groups, especially expounding to an amphitheater full of students. Both systems of education depended upon abilities and efforts of individual practitioner-doctors, who were the main attraction. In general, throughout this period medical schools played a peripheral role, operating through the doctors who simultaneously occupied positions on their faculties and on the staffs of the city hospitals.

# 4 · Comfort and Care: The Ongoing Struggle

> A hospital for the poorest of the poor, the
> dregs of society, the semi-criminal, starving,
> unwelcome class, who suffer and die unrec-
> ognized, and to whom charity at best is
> cold and mechanical...[T]here is no luxury
> here; not much gentleness.
> Description of Bellevue Hospital, 1878

IN THE WINTER of 1864 a penniless songwriter,
weakened from three days of chills and fever, stumbled and fell in his
attic room. A friend, finding him on the floor bruised and bleeding,
took him to Bellevue Hospital, where he was put to bed. When the
friend returned, the composer said that nothing had been done for
him and he couldn't eat the food they brought him. The next day he
was dead; only another visit by his benefactor saved Stephen Foster
from burial in an unmarked grave.[1] Had he lived he might justifiably
have written a sequel to "Old Folks at Home" called "Home Was
Never Like This." Yet each year thousands of poor in the big cities of
the United States sought the scant attention and scantier food of city
hospitals as the only refuge from unrelieved suffering and ignomini-
ous death in a garret room or crowded tenement.

Like many another, Stephen Foster entered a city hospital as an
emergency patient. Seriously ill patients had been dumped at the
doors of city almshouses in the past; their number increased as the
almshouses turned into hospitals. Other patients in Foster's ward
would have been admitted from the dispensary or sent in by private
physicians, still others brought in by police because they were
considered public nuisances—alcoholics, drug addicts, or starving,
wasted derelicts. Finally, some would have been wards of the local
government, confined in jails, almshouses, or orphans' homes until
illness forced their transfer to the public institution for the sick.

After this flotsam and jetsam was settled into beds in the city
hospital, how was it cared for? At first, medical and personal care was
little better than during the preceding almshouse period. The doctor's
understanding of many illnesses was vague (see Chapter 1). For
instance, at the Boston City Hospital in 1867 the most frequent
diagnoses were cough, constipation, bronchitis, debility, and pain.
Really effective remedies were few: doctors had opiates for pain,
digitalis for heart failure, quinine as a specific remedy for malaria, and

little else. Doctors could perform external and minor operations and set fractures but dared not venture farther for fear of infection, which was all too common following the operations they did perform. "Everything swam in pus," an intern at Boston City Hospital later wrote of conditions in 1877. "We even had at one time so much hospital gangrene that all operations were suspended for several weeks, because to cut a man meant to kill him.[2] Although infections were found in all hospitals, public and private, they tended to spread more widely in public hospitals because patients were usually more debilitated, conditions were less conducive to cleanliness, and crowding was more common than in private hospitals.

Because patients in city hospitals tended to have advanced illnesses, because so little could be done except wait for death or recovery, and because so few had suitable homes to return to, most patients remained in the hospital a long time. At the Philadelphia almshouse in 1807 the length of stay for bedridden patients averaged a year in the general wards and three to seven years in wards for the incurables. Turnover had accelerated little, if any, by 1877, when on a typical day a doctor at the Philadelphia Hospital noted that many of the patients were "convalescent...out in the yard sitting, smoking, or reading." Two decades later the chief resident could still complain that the wards were filled with patients who needed little or no medical treatment.[3] The abundance of persons requiring long-term care on hospital wards was one reason why it proved so difficult to loosen the ties that bound hospitals and almshouses.

The less the doctors could achieve, and the longer patients stayed in the hospital, the more important became other factors, such as good food and clean surroundings, especially in city hospitals, where many patients needed to recover from the effects of squalor and near-starvation preceding admission. Whether their doctors were skillful and conscientious or inexperienced and careless, for many patients recovery depended more on expert nursing. But nursing was made more difficult because the city hospital required the nurse to do double duty—to minister to patients personally and to supervise a motley crew of almshouse inmates, convalescent patients, and ill-trained attendants who were assigned to keep the wards in order and attend the sick. Under these circumstances, how good was the nursing?

In the mid-nineteenth century, when nurses were still considered little more than servants, few were attracted to nursing as a career. The handful who became respected ward supervisors had to work for years to attain that status, since experience was their only teacher. Those with the ambition and ability to become supervisors tended to choose voluntary hospitals, where their talents were more likely to be recognized and honored, than to labor in a morass of dirt, disorder, incompetence, and hopelessness, where political pull brought advancement faster than ability. Thus, although the better private hospitals were able to select suitable candidates and give them good on-the-job training, many head nurses in city hospitals were either lazy or dull, indifferent or cruel, or a combination of these.[4] Occasionally a political appointee or a convalescent patient with intelligence, industry, and native administrative skills would through long years of hard work reach the position of chief nurse of a ward. In the 1880s at the Philadelphia Hospital two such nurses—one in charge of the men's medical and the other of the men's surgical ward—were highly praised by house officers and attending doctors for their personal qualities as well as professional skills. Osler referred to one of them as "dear old Owens...with his Hippocratic gift of prognosis."[5]

In many city hospitals a corps of almshouse inmates, petty criminals, and convalescent patients, both men and women, stayed on after discharge to work for a pittance as helpers. Even if they had had the wit to do so, without formal instruction they would not have known how to detect significant changes in a patient's illness or how to respond to them if detected. Most were content to run errands, clean and scrub after a fashion, and perform other menial tasks.

A similar atmosphere of ignorance and futility also pervaded city hospitals that had not begun as almshouses. As an assistant nurse at the Boston City Hospital in 1872, Linda Richards, who later became director of nursing at Massachusetts General and Boston City Hospitals, discovered that she was to be taught nothing about patients' symptoms nor was she expected to pay any attention to them. She found most of the senior nurses to be "thoughtless, careless and often heartless...[They] were not respected, most of them being called by their first names by the doctors and by everyone in the hospital... [N]o one seemed to have any supervision over them or care in the least what their conduct was," a prescription for mediocrity or worse.[6] Under these conditions seriously ill patients would receive proper nursing care only if they chanced to be on a ward run by a conscientious nurse who could spare a minute here and there for bedside nursing. Sometimes a medical student or intern might nurse a desperately ill patient through a critical period. An intern at the Philadelphia Hospital in the 1880s recalled thirty years later: "This absence of trained nurses threw much more responsibility on the internes than is now the case in any hospital. We did all the surgical dressings, and were forced to oversee very closely the details of the nursing itself."[7]

The deplorable conditions within hospitals were sounding alarms in many places; thoughtful doctors and laymen alike were becoming concerned. The United States and Great Britain glimpsed the deficiencies of hospital care during the Civil War and the Crimean War. Following her experiences in army hospitals during the latter, Florence Nightingale opened a school of nursing at St. Thomas's Hospital, London, in 1860. This school, in which hand-picked candidates were given three years of training in nursing, became a model for others all over the world and especially in the United States. The Nightingale system emphasized training secular nurses in such a way as to attract superior women to the profession, offering education in medical subjects, and securing hospital conditions that made these objectives possible, including female supervision of the nursing services.[8]

In New York City a new era of reform was evolving, marked by the exposure of Boss Tweed and the corrupt Tweed Ring in the 1870s. In 1872 Louisa Schuyler, a leader in the formation of the Sanitary Commission, which had rendered outstanding service in hospitals during the Civil War, called together a group of society women. Incensed at what they heard about Bellevue Hospital and imbued with the humanitarian spirit of the times, the ladies formed a committee and left "the moistureless froth of the social show" long enough to investigate the hospital. One of them, Elizabeth C. Hobson, wrote of her shock as she pried and searched where few society matrons had ventured before.[9] She found that "the condition of the patients and the beds was unspeakable; the one nurse slept in the bath-room, and the tub was filled with filthy rubbish. As for the nurse,...to her was confided the care of twenty patients, her only assistants being paupers,...women drafted from the workhouse, many of whom had been sent there for intemperance, and those convalescents who could leave their beds."[10] One of the doctors, W. Gill Wylie, added, "There were no night-nurses; the night-watchmen—three in number to a hospital of eight hundred beds—were expected to give assistance to patients requiring attention during the night."[11] The committee, with the help of several concerned doctors, established the New York Training School for Nurses and persuaded the authorities of Bellevue to delegate to it the responsibility for nursing in the hospital. With London-trained Helen Bowden as the first superintendent, the school opened in 1873, the first in the United States to be organized on the Nightingale plan.[12] Similar schools were started later the same year at the Massachusetts General and Connecticut State Hospitals, and schools associated with other city hospitals opened soon after at Charity Hospital in New York City in 1875 and at the Boston City Hospital in 1877 or 1878.[13]

Chicago followed New York's lead closely. In 1880 several public-

spirited women, who were familiar with training schools in England
and the eastern United States, organized the Illinois Training School
for Nurses in conjunction with Cook County Hospital. Because the
warden was hostile and the county commissioners indifferent, the
school had to be content with control of two wards at first, but it
rapidly expanded throughout the hospital as staff and administration
came to appreciate the skills and devotion of the nurses. Sometimes
their diligence may have been excessive: by 11:00 A.M. all routine care
of patients had to be finished and the beds in every ward lined up "so
that all brass knobs were in an undeviating straight line." The school's
progress was perhaps measured better by its educational program. In
1896 it adopted a three-year curriculum, well in advance of many
schools in the country. Postgraduate courses were offered as early as
1899, and in 1905 the Illinois school became the first large school to
offer affiliations with smaller ones, enabling them to meet requirements for state registration.[14]

In an environment so highly charged with politics as was the Cook
County Hospital, the leaders of the Illinois Training School struggled
to stay clear of political entanglements. Attempts were made to
incriminate the school by false charges. For instance, in 1887, when
the school was being paid $100 per month per ward to supply nursing
services, it was accused of collecting for a ward that had been closed
for some time. The school's leaders were able to show that in fact they
had provided gratuitous nursing when this ward had been opened
periodically for overflow patients from a nearby ward. Several years
later Cook County authorities publicly accused the nursing school of
keeping patients on the wards longer than necessary in order to get
work out of them. The school easily refuted this absurd charge by
pointing out that patients were retained or discharged only on doctor's
orders. This attack ceased, but, according to the custom of politicians,
no apology was forthcoming. The reason for the harassment was
obvious to the initiated: Cook County authorities used this maneuver
to throw the opposition off balance when the yearly contract between
hospital and nursing school was being renegotiated. Similarly, a few
years previously the county commissioners had threatened to substitute a hospital school of nursing for the Illinois Training School and
then had withdrawn the threat after the annual contract was agreed
upon. Such are the ways of politicians.

It is not surprising, in view of such shenanigans, that the governing
board of the Illinois Training School, having built up the finest nursing
school in the Midwest and one of the best in the United States,
welcomed the overtures of the University of Chicago, which wanted
to convert the school into the college of nursing of the university and

the department of nursing of the university hospital. The school was moved in 1929, at which time the Cook County School of Nursing was founded to replace it. This school adopted the mission, program, and ideals of the original school, including the principle of dual authority, a board of trustees that ran the nursing school and the nursing service of the hospital, and administrators appointed by the county commissioners, who ran the rest of the hospital.[15]

The movement for women to sponsor nursing schools in affiliation with city hospitals and to dominate their governing boards coincided with strong female interest in civic causes and vigorous support for women's rights. This arrangement gave the hospital a voice in high places where its needs were seldom publicized, and women on the school's governing board and committees used their social contacts assiduously to obtain needed funds and favors. The device of establishing a nursing school separate from but interlocking with a city hospital was also partially successful in keeping the school free of political entanglements. But the system also had its disadvantages. By ceding to women the nurses' training school as their territory, politicians and administrators managed to insulate them from the other activities of the hospital. By tacit agreement those who ran the nursing school understood that in return for immunity from interference the school was expected to refrain from criticizing the hospital. Yet their destiny, like that of the rest of the hospital, was decided in smoke-filled rooms; by controlling the budget of the nursing school, the politicians in effect controlled the school. As usual, the power of the purse was the power to rule.

In retrospect, although semi-independence was helpful in protecting some schools until their survival was assured, in the long run a more direct relationship between school and hospital proved a better system. Such was the arrangement followed from the beginning at the Boston City Hospital. When the nursing school was established, it was the fifth in an American hospital and the first of these to be controlled directly by the board of trustees of the hospital. The superintendent of nurses reported to the superintendent of the hospital, as did other department heads. Despite the incursions of Boston politics elsewhere in the city hospital, there was no evidence of political interference in the nursing school.[16]

At the Philadelphia Hospital the organization of a nurses' training school was held up for over a year because politically-minded members of the governing board objected to the appointment of a director of nursing from outside. Finally, some leading citizens of the city took matters in their own hands and obtained a commitment from Alice Fisher, chief of nursing at Birmingham Hospital, England, who was recommended by Florence Nightingale. After two of these citizens offered to pay part of her salary, the board agreed to appoint her, and the school opened in 1884.[17]

The nursing school at Charity Hospital in New Orleans was also resisted initially. Beginning in 1881 or before, attempts to open a nursing school were blocked by friends of the Sisters of Charity, who had been installed in the hospital in 1834 to bring order out of administrative chaos. Their supporters, who feared the training school was a scheme to displace the sisters, yielded when the sisters consented to the establishment of a school under their direction. This was organized in 1894.[18]

In Washington the city hospital was at first a minor affiliate in the development by other hospitals of a nursing school, though it eventually came to control the school. The Washington Training School for Nurses, organized in 1877, gave instruction at first at Columbia Hospital for Women and Freedmen's Hospital, adding Emergency and Garfield Hospitals between 1883 and 1885 and the Washington Asylum Hospital in 1893 or 1894. In 1898 the attending physician at the asylum hospital reported that pupil nurses from the training school had replaced "incompetent nurses" in the wards. By 1901 only Emergency and Washington Asylum Hospitals remained as affiliates of the school. Three years later the school was reorganized as the Capital City School of Nursing and placed within the nursing division of the asylum hospital.[19]

Thus, by the early twentieth century many city hospitals had successfully launched nursing schools. But Bay View Asylum in Baltimore failed in its attempt. The hospital opened a school in 1911, but within a few years the chiefs of services complained that the hospital's student nurses were being barred from the state's examinations because of inadequate facilities at the nursing school. When funds were not forthcoming to remedy the defects, the hospital gave up trying to train registered nurses. In 1925 a school for practical nurses was opened. This satisfied some nursing needs, but the lack of a full-fledged school hampered the recruiting of sufficient registered nurses to staff the hospital properly.[20]

City hospitals were often the first, or among the first, in their localities to establish nurses' training schools. For instance, Osler wrote of Alice Fisher: "The good work which she has accomplished has stimulated other hospitals of the city, and training-schools have been established at the Pennsylvania, Episcopal and University Hospitals."[21] To be sure, city hospitals were usually larger and offered more varied facilities than private hospitals, but there was a more compelling reason for them to pioneer: they required more personnel per patient than other hospitals because their patients were sicker and their antiquated, worn-out equipment was less efficient. Yet, they had difficulty recruiting nurses. Hospital administrators saw nursing students as a continuous source of cheap help. They may or may not have anticipated also that trained supervisory nurses would provide a cadre of reliable administrators for wards and outpatient clinics, a resource especially needed when so few doctors worked full time at city hospitals.

Directors of nursing schools knew that antagonism toward their administration existed, so they made haste slowly. They appointed responsible trained nurses to key positions as instructors and ward supervisors while keeping the most reliable of the nurses who had had no formal training. Alice Fisher wrote the president of the Board of Guardians of the Philadelphia Hospital: "By all means retain all nurses who are honest and of good character; they can always be made useful, and if not offended may be very valuable." But still some were offended; almshouse inmates who had been removed from nursing jobs threw rotten eggs through the director's window in protest. Nevertheless, within two years Fisher managed to revolutionize nursing at the Philadelphia Hospital.[22]

The story was the same in the schools at other city hospitals. Recruits were hand picked and generally highly dedicated. Sometimes they had to muster all the devotion they could, as at the Boston City Hospital, where student nurses slept four to a room designed for one until a bequest provided for a well-appointed nurses' home, opened in 1899. Since the Nightingale system called for a strictly supervised regimen for student nurses, both on and off the wards, officials recognized the need for a separate residence building with space for recreation as well as classrooms, and cities without ready donors— which was most of them—obtained appropriations for nurses' homes as soon as they could.[23]

Nurse-administrators had to be models of diplomacy to get along with parochial, politically-minded, recalcitrant governing boards and with staff doctors loath to change their routines. But as these and other adversities were overcome, the nursing schools demonstrated their potential. Yet, even the best nursing administrator could not develop good nursing services without proper support. City hospitals seemed never to have enough nurses on the payroll. Salaries were low because the poorhouse philosophy prevailed almost everywhere, and city hospitals suffered in competition with private hospitals in recruiting the best nurses. In 1893, two decades after the nursing school had opened at Bellevue, a typhus epidemic revealed that the hospital was regularly using convalescent patients as ward attendants. As late as 1900 repeated complaints of poor service and cruelty to patients forced the commissioner of hospitals to discharge half of the unpaid help, and a few months later an investigation revealed that "with a few notable exceptions, inefficiency, callousness, and idleness characterize those who are employed to do the work of the hospital."[24]

Satisfactory nursing was hampered by other factors such as poor food, crowding, dilapidated buildings, and outworn equipment. Food in city hospitals was never abundant and was one of the first items on which administrators skimped when the always low budget sank even lower. It was, moreover, poorly cooked and carelessly served. One explanation why such conditions were tolerated was given by the trustees of the Boston City Hospital in 1884, when doctors at the hospital complained that the patients' food was scanty and unappetizing, even disgusting. The trustees replied that the food served in the hospital was what all people of average means should expect.[25]

At Bellevue in 1882 meals consisted of one slice of "damp and unpalatable" bread and one cup of coffee plus one teaspoonful of very diluted condensed milk for breakfast and a slice of bread and a cup of tea without milk or sugar for supper. The only other food was a little oatmeal gruel in midmorning and a "small cut of overcooked beef" and a slice of bread for dinner. If the patients were not undernourished when they arrived, they were sure to be after few weeks in the hospital. Around this time the diet in the Philadelphia Hospital was such that some patients actually developed scurvy. In the year 1883–84 the daily cost of food for patients and employees at Cook County Hospital was 24.3 cents per person (equivalent to $1.10 worth of food in 1967).[26]

During the same year the total daily expenditure per patient at that hospital was $1.01, which included medical and nursing care, food, supplies, purchase and upkeep of equipment, maintenance of buildings, and administrative overhead. Conditions were slightly better at the Boston City Hospital, where the total daily expense in that year was $1.24 per patient (comparable to $5.65 a day in 1967). In 1915, when the total daily expenditure per patient at Cook County Hospital had increased slightly to $1.40, barely keeping up with the increased cost of living, it was only 72 cents per day in the general hospital of the Bay View Asylum in Baltimore. Amounts spent to care for ward patients in private hospitals and the city's municipal hospital were contrasted by the Board of Charities of the District of Columbia in 1901: $1.01 to $1.29 per day in private hospitals (except for 46 cents per patient per day in a Catholic hospital, where much of the nursing was done by sisters, for whom presumably no charge was made) versus 34 cents a day at the Washington Asylum Hospital. (See Table 5 for other comparisons of costs at private and public hospitals.) We can reasonably assume that along with higher costs went extra comforts

Table 5. Comparison of costs for patients at public and private hospitals, selected years.

| Year | City | Public hospitals | | Private hospitals[a] | |
|---|---|---|---|---|---|
| | | Hospital | Cost/patient/day | Hospital | Cost/patient/day |
| 1883 | Boston | Boston City | $1.24 | Massachusetts General | $1.80[b] |
| | Chicago | Cook County | 1.01 | — | — |
| | New York | Bellevue | 0.49 | Mount Sinai | 1.03 |
| 1904 | Boston | Boston City | 1.55 | Massachusetts General | 2.42[b] |
| | New York | Bellevue | 1.25 | Mount Sinai | 2.51 |
| | Washington, D.C. | Washington Asylum | 0.58[c] | Garfield Memorial | 1.50 |
| 1915 | Baltimore | Bay View Asylum | 0.72[c] | Johns Hopkins | |
| | | | | All patients | 3.47 |
| | | | | Private patients | 4.70 |
| | Boston | Boston City | 2.34 | Massachusetts General | 3.31[b] |
| | Chicago | Cook County | 1.40 | Presbyterian | 2.97 |
| | New York | Bellevue | 1.63 | Presbyterian | |
| | | | | Ward patients | 3.19 |
| | | | | Private patients | 7.99 |
| | Washington, D.C. | Washington Asylum | 1.08[c] | Garfield Memorial | 1.65 |
| | | | | George Washington University | 2.16 |

*Sources:* Figures for 1883 and 1915 are from the annual reports of the individual hospitals. Those for 1904 are from J. N. E. Brown, "The Per Capita Cost," *National Hospital Record* 10 (January 1907):16–18, except the figures for Washington, D.C., which are from annual reports.

a. Unless otherwise stated, it must be assumed that these figures were for all patients—in private rooms as well as on the wards.

b. In 1883 the number of private patients at the Massachusetts General Hospital varied from 3 to 8. In 1904 only 1.2 percent of patients were private patients, and in 1915, 3.6 percent. The Massachusetts General Hospital established its first pavilion for private patients in 1917. See Trustees of the Massachusetts General Hospital, *Annual Reports, 1883, 1904, 1915*; Morris J. Vogel, *The Invention of the Modern Hospital: Boston, 1870–1930* (Chicago: University of Chicago Press, 1980), p. 114.

c. Hospital patients only.

and better care. For one reason, voluntary hospitals contained some private patients from the beginning. Furthermore, the indigent patients in those hospitals were selected, tending to come from the servant and artisan classes, and were often recommended by one of the hospital's patrons. Consequently, wards were likely to be cleaner, the quality of help better, and nursing more often adequate.[27] At least some progress was made in food service in city hospitals: in 1909 authorities at Cook County Hospital reported that rather than using convalescent patients, they were using carts manned by hospital personnel to serve meals, and in 1915 crockery was substituted for tin cups on patients' trays. Tin cups were abandoned in Baltimore's city hospital four years later.[28]

Another handicap to good patient care was crowding. City hospitals grew rapidly between the Civil War and the First World War, in both the older and the newer cities. The bed capacity of the Boston City Hospital increased from 230 in 1870 to over a thousand in 1910, admissions from 2,569 to 14,442 per year. Cook County Hospital contained 220 beds in 1870 and 1,350 in 1910; admissions increased from approximately fifteen hundred to nearly thirty thousand.[29] Keeping up with the numbers of patients who absolutely had to be admitted was a continual problem. It seemed that these hospitals were always planning, asking for funds, or floating a bond issue for a new building. Actually, the use and reuse of old buildings was more common than the erection of new ones. Some were so dilapidated and outdated that they literally came down about the heads of patients before they were abandoned (see Chapter 2, n.35). They were used because they were at hand and funds for new ones were not, because cheap poorhouse and convalescent labor could be used to make repairs, and because paupers could not be choosers. But despite the construction of some new buildings and the reuse of many old ones, crowding persisted. City hospitals had not built a better mousetrap, but for many persons they offered the only mousetrap available. In 1873 the Philadelphia Hospital, with a capacity of five hundred beds, contained a thousand patients, many of them "on beds laid about on the floor and in every available corner." Doctors in these wards were "almost driven out of them by the stench" of so many patients, poorly cared for, with woefully inadequate facilities for disposal of garbage and excreta. At Charity Hospital in New Orleans patients were often kept two to a bed as late as 1914.[30]

Patched-up buildings were put into use, and new buildings were erected when funds could be obtained from reluctant legislators and an equally reluctant electorate. Under the circumstances, city hospitals became architectural nightmares. Sometimes they were touted by city officials as showpieces, but even then all they could boast about was their immensity. The desire for vastness, and the publicity, power, and opportunities for graft that went with it, so overcame the politicians of Cook County in 1911 that they planned a hospital of four thousand beds, a city block piled full of cubes and rectangles ten or more stories high. Dr. Sigismund S. Goldwater, a distinguished New York hospital administrator who was consulted, was appalled at the idea of "five or six miles of sickbeds under one management, an ungovernable mass which spells outrage and disaster." Hospital authorities in other cities generally followed Goldwater's advice, but in Chicago authorities went ahead with their grandiose plans, and the Cook County Hospital of the 1950s fulfilled his prophecy. To the observer, it appeared that cupidity mated with stupidity had produced a monstrosity.[31]

In contrast, New York City developed early an overall plan for its municipally-owned hospitals. By 1898 the sections of the city not supplied with private hospitals had three municipal hospitals: Gouverneur, Harlem, and Fordham. Patients needing longer, more sophisticated medical care were sent, from these hospitals or directly, to Bellevue or City Hospital, adults with chronic illnesses to City or Metropolitan Hospital, and chronically ill children to the Children's Hospitals or the Infant's Hospital. Two nursing schools supplied the general hospitals; one was at Bellevue and the other was related to City Hospital and the three reception hospitals.[32]

As the practice of medicine became more scientific, and as it called more and more upon technology for help in diagnosis and treatment, the equipment required in a hospital became increasingly complex and expensive. This need became acute early in the twentieth century. A good example was the apparatus necessary for X-ray examinations. X rays had been discovered by Wilhelm Conrad Roentgen in November 1895 and almost immediately applied to the diagnosis of disease. One of the first X-ray units was established in 1896 at the Boston City Hospital by Francis Henry Williams, and by 1899 a department composed of one X-ray machine, one technician, and one doctor was operating at the Philadelphia Hospital. Williams and other pioneers soon demonstrated the value of this new diagnostic method, and the larger hospitals hastened to install X-ray apparatus. But this equipment was expensive to build, became progressively more elaborate as X-ray examinations became more extensive, and, like other products of a fast-developing technology, rapidly became obsolete and had to be replaced. The steeply mounting costs were too much for the meager budgets of most city hospitals. For instance, the X-ray department at Cook County Hospital before 1913 "struggled along as best it could in a tiny, cramped space" containing "antiquated coils, wornout X-ray tubes, and a few articles which were in a musty or completely unusable condition. Nothing but fracture cases and a few bone lesions were examined because of the lack of proper equipment." Only by renting apparatus from a local manufacturer was the department able to widen its scope a little. With the opening of a new building in 1914, the X-ray department was given space, equipment, and staff that were adequate, at least for a while.[33]

This was the typical history of X-ray and other facilities in city hospitals: shortages followed by famine, then relief, and then the whole cycle starting over again. Barely scraping along from day to day, city hospitals had nothing left for replacement costs. All kinds of equipment—from beds and mattresses to microscopes and operating tables—wore out or became obsolete long before replacements could be bought. Wear and obsolescence galloped beyond reach while the budget limped behind, and the distance increased as the pace of technology accelerated. Worse still, some hospitals did not even try to cope (see Chapter 5 regarding deficiencies of equipment at Baltimore City Hospitals).

Doctors, nurses, and attendants who were forced to scrimp and scrabble to provide for the barest physical and medical needs of their patients had little time or energy in reserve for ministering to patients' emotional needs. In the absence of specific remedies, and while sickness was still considered a visitation for sin, hospitalization had been regarded as an opportunity for moral uplift. The Children's Hospital of Boston stressed its spiritual role and attempted to "cultivate the devotional feelings" of its patients.[34] Other private hospitals struck the same note, and the Chicago Hospital for Women and Children was founded to provide a place for those who would otherwise be exposed to the coarseness and impersonality of a public ward at Cook County Hospital.[35] Even the most hypocritical politician would find it difficult to claim that patients in noisy, smelly, dirty, crowded, vermin-ridden wards of the city hospital, attended by careless fellow patients, harried nurses, and overworked doctors, could expect much spiritual elevation. When the study of pathology taught doctors to pinpoint the diagnosis of disease to its location in individual organs, when safer surgery and effective medicines confirmed the value of an organic approach to disease, interest in the patients' emotional needs receded into the background, even in voluntary hospitals.

By the early twentieth century, however, a reaction had set in. Voices rose in protest against the exclusive focus on organic disease and the mechanization of the hospital routine. "It broke my heart all the time...when the nurse folded sheets," a patient complained to Jane Addams; she had desperately needed the nurse's attention herself. Out of the protests came the installation of social service departments in hospitals; the first in the United States was instituted at the Massachusetts General Hospital in 1905, with other private hospitals following. These departments helped make hospital and community services available to patients and their families, counseled them, and helped them adjust emotionally to illness while continuing to minister to their social and emotional needs after they left the hospital.[36]

In city hospitals, where patients had empty pockets and few friends and often lacked the skills needed for daily living in a big city, social workers were needed even more. In 1906 or 1907 Bellevue appointed a social work nurse whose duty it was to help patients who were about to be discharged. Cook County Hospital's social services began in 1911, when social workers were installed to provide for the future of unmarried and homeless mothers.[37] Although the need for social workers at the Boston City Hospital was greater than at the Massachusetts General, no department was started until 1916. A committee of private citizens assumed the financial responsibility for a department until it could be worked into the city budget, and Gertrude Farmer was brought from the Massachusetts General Hospital to direct it. Three years later the Women's Advisory Council of the Philadelphia Hospital adopted the same strategy to start a social service department there. The department at Charity Hospital in New Orleans began in 1915, and it too was supported by donations until the hospital took over in 1921.[38]

Throughout the nineteenth century the public's image of most city hospitals continued to be marred by the stigma of the almshouse, the impression that only persons from the scum of society—the loafer, the ne'er-do-well, and the criminal—would consent to enter them. Despite all the evidence that honest, industrious men and women could be out of work through no fault of their own, the attitude prevailed that abject poverty was caused either by laziness or wickedness and was in any case sinful; thus, almshouse hospitals were thought of as houses of refuge for the immoral. The Boston City Hospital tried, though with only partial success, to aim higher from the beginning. It was not to be "a hospital for the reception of the degraded victims of vice and intemperance, or a home for the helpless pauper"; rather its purpose was to alleviate the sufferings and restore the health of those who were temporarily laid low by illness. The hospital's rules soon specified that persons with acute venereal disease or alcoholism should not be admitted unless they were paying patients. These bans proved unworkable, however. Alcoholics who landed on the hospital's doorstep had to be admitted if they were suffering from other illnesses; and after Paul Ehrlich announced in 1907 that arsphenamine would cure syphilis, patients with acute syphilis were admitted to private as well as public hospitals.[39]

Another action of the trustees helped raise the status of the Boston City Hospital: the provision of private rooms. A few were included in one of the original buildings, enabling staff physicians to admit a private patient if such action were justified. Originally, such a patient might have been a well-to-do traveler with no relatives in Boston to whose home he could repair in a time of illness; later, these rooms were used also for politicians and their friends, who expected and got excellent care and extra favors, and for a patient whose illness was of special interest to one of the staff doctors. When the hospital was crowded, the rooms were used for nonpaying patients. An inquiry in 1868 revealed that, of the 4,838 patients admitted to the hospital, thirty-two had compensated their doctors either by a gift or by paying a fee. None of them were residents of Boston. While accepting these fees was not expressly permitted, the trustees allowed it, and in 1906 they legitimized the custom. By that time most leading private hospitals had already adopted the practice, and the Massachusetts General Hospital was planning facilities for private patients.[40]

From the opening of the Boston City Hospital, patients who could pay for their board and room were expected to do so, and the authorities soon began to collect from towns elsewhere in Massachusetts when their citizens entered the hospital. By 1878 the trustees declared that the public was beginning to understand that the institution was "not a free hospital, but a place where it is right and proper to pay, and where all must pay what they can for the good they get." In 1907 the editor of the *Boston Medical and Surgical Journal* announced that nearly one-fifth of the $500,000 needed to run the city hospital had been collected from patients, a "definite step toward the solution of the problem of medical charity." Whatever the reason, it is estimated that the percentage of white-collar workers among patients in the Boston City Hospital increased from 10.5 to 19.6 between 1880 and 1900, while the percentage of blue-collar workers in the total decreased from 36.1 to 24.1.[41]

Hospitals that started as almshouses abandoned slowly and reluctantly their practice of admitting only the indigent. In Washington, for example, as long as the city hospital was housed in old buildings and bore the name of Washington Asylum Hospital, the patients were considered paupers. Only in 1924, after the name was changed to Gallinger Municipal Hospital and the first buildings were erected specifically for hospital use, were the authorities allowed to charge psychiatric and tuberculous patients for room and board. Since these were patients who were seldom admitted to private general hospitals, opposition to such charges could be overcome; paying patients were not accepted throughout the hospital until 1943.[42]

The provision of a few private rooms at the Boston City Hospital and the admission of paying patients at the city hospitals of Boston and Washington elevated the status of the hospitals somewhat in the eyes of the public and, in the case of the Boston hospital, undoubtedly helped attract gifts. But much of the spirit of the pauper hospital remained—the willingness to make do, to provide third-rate care because that was all patients in these hospitals were supposedly accustomed to and thus all they could expect. Administrators and politicians sensed but did not confront the basic issue: third-rate care was all the public was willing to pay for.

Yet, despite the persistent pinch-penny philosphy, progress was made between 1860 and 1910, resulting especially from three forces: the rapid growth of medical science, improvements in medical education, and the professionalization of nursing. City hospitals participated in all three movements and benefited from all three. At the beginning of the period the city hospital, as James R. Wood said of Bellevue, "was comparatively without scientific resources, without organization, and without influence," a hospital struggling with a small staff and little professional help to take care of acutely ill patients housed along with the aged, decrepit, handicapped, and homeless.[43] By the end of the period the older city hospitals had become independent of the poorhouse and the insane asylum. Although they were always straining to catch up with the sweeping changes in medical care and in medical and nursing education, their staffs managed to become professionalized. The training and education of doctors, nurses, and administrators improved strikingly, and these members of the medical team acted more as a group in 1910 than they had in 1860. Professionalized superintendents and nurses, active attending staffs, organized house staffs, and new, comprehensive programs for educating medical students all contributed to a spirit of making these hospitals work in spite of handicaps. The city hospitals were poised to take advantage of the astounding progress in medical care that the twentieth century would bring.

# Part III · *The Academic Period*

# 5 · *Cities and Their Hospitals: An Uneasy Equilibrium*

> It was overcrowded, understaffed, its ac-
> creditation had almost been withdrawn...;
> but it was probably no more backward,
> politics ridden or neglected than similar in-
> stitutions of charity all over the country.
>
> Jan de Hartog

> The vast majority [of city hospitals] are
> badly run, impoverished, long-neglected
> fleabags.
>
> Lewis Thomas

AFTER THE FIRST World War continuing immigra-
tion and increasing rural-urban migration kept American cities
growing. Between 1910 and 1960 the population of the United States
increased from nearly ninety-two million to 178.5 million. At the
beginning of this period 46 percent of the people lived in urban areas;
fifty years later the figure was nearly 70 percent. The number of cities
with more than one hundred thousand inhabitants rose from fifty to
132, and five passed the million mark. The larger cities shared in the
nation's growth. Washington more than doubled its population
between 1910 and 1960; Baltimore, Chicago, New Orleans, and New
York increased theirs by about two-thirds and Philadelphia by
approximately one-third (see Table 6). Boston's seeming stagnation
was an illusion. Because of the city's fixed boundaries, growth merely
pushed people into the suburbs.[1]

Equally significant were the kinds of people who tended to occupy
urban areas. With few exceptions, the business of cities was business
(to borrow President Coolidge's phrase). As America flourished,
business and industry reached out for more room. As they preempted
much of the area within the city boundaries, many people fled to the
suburbs, which promised space, freedom, and lower taxes. Left behind
were those unable or unwilling to move, especially the poor, the aged,
the infirm, the ill-adjusted, and the poorly trained. Since these groups
had more than their share of sickness, hospitals were badly needed;
and for the many who could not pay for care, city hospitals were the
principal refuge. By the mid-twentieth century private hospitals had
joined the exodus to the suburbs, and the burden on the public
hospitals became even greater.

Many of those who remained in the cities were immigrants. In 1910
about two-fifths of the population of Boston, Chicago, and New York

Table 6. Population of selected cities by race and nativity, 1910, 1930, and 1960.

| City | Population (= 100%; in thousands) | | | Black (%) | | | Foreign-born white (%) | | | Native white, foreign parentage (%) | | | White, native parentage (%) | | |
|---|---|---|---|---|---|---|---|---|---|---|---|---|---|---|---|
| | 1910 | 1930 | 1960 | 1910 | 1930 | 1960 | 1910 | 1930 | 1960 | 1910 | 1930 | 1960 | 1910 | 1930 | 1960 |
| Baltimore | 558 | 805 | 939 | 15 | 18 | 35 | 14 | 9 | 4 | 24 | 20 | 11 | 47 | 53 | 50 |
| Boston | 671 | 781 | 697 | 2 | 3 | 9 | 36 | 30 | 16 | 38 | 42 | 30 | 24 | 25 | 45 |
| Chicago | 2,185 | 3,376 | 3,550 | 2 | 7 | 23 | 36 | 25 | 12 | 42 | 40 | 24 | 20 | 28 | 41 |
| New Orleans | 339 | 459 | 628 | 26 | 29 | 38 | 8 | 4 | 2 | 22 | 14 | 6 | 44 | 53 | 54 |
| New York | 4,767 | 6,930 | 7,782 | 2 | 5 | 15 | 40 | 33 | 20 | 38 | 40 | 29 | 20 | 22 | 36 |
| Philadelphia | 1,549 | 1,951 | 2,003 | 5 | 11 | 27 | 25 | 19 | 9 | 32 | 32 | 20 | 38 | 38 | 44 |
| Washington | 331 | 487 | 764 | 28 | 27 | 54 | 7 | 6 | 5 | 14 | 13 | 8 | 51 | 54 | 33 |

*Sources:* For 1910: U.S. Department of Commerce, Bureau of Foreign and Domestic Commerce, *Statistical Abstract of the United States, 1924* (Washington, D.C.: U.S. Government Printing Office, 1925), pp. 42–45; U.S. Department of Commerce, Bureau of the Census, *Negro Population, 1790–1915* (1918; reprint ed., New York: Kraus Reprint, 1969), p. 93. For 1930: U.S. Department of Commerce, Bureau of the Census, *Statistical Abstract of the United States, 1939* (Washington, D.C.: U.S. Government Printing Office, 1940), pp. 20–25; U.S. Department of Commerce, Bureau of the Census, *Negroes in the United States, 1920–32* (1935; reprint ed., New York: Kraus Reprint, 1969), pp. 54–55. For 1960: U.S. Department of Commerce, Bureau of the Census, *County and City Data Book, 1962: A Statistical Abstract Supplement* (Washington, D.C.: U.S. Government Printing Office, 1962), pp. 486, 496, 506, 516, 536, 556.

was foreign born and another two-fifths were children of foreign born (see Table 6). By 1930 the pace of immigration had slowed a little: one-third of New York's population was foreign born, as was three-tenths of Boston's and one-fourth of Chicago's. Two-fifths of the people in these three cities were children of immigrants. Even as late as 1960, immigrants made up one-fifth of New York's population, one-sixth of Boston's, and one-eighth of Chicago's, while children of immigrants made up from 24 to 30 percent of the population of these three cities.

Meanwhile, a black tide had flowed in from the South. Washington, D.C., had attracted large numbers of blacks ever since the Civil War, and in 1910 the population of the capital city was 28 percent black, a greater proportion than in the nearby border city of Baltimore (15 percent) and the southern city of New Orleans (26 percent black). Northern cities had very small black populations at this time: only 2 percent in Boston, Chicago, and New York and 5 percent in Philadelphia. Since the black migration to the North came much later than the foreign immigration, these percentages had not increased significantly by 1930 (see Table 6); but by 1960 the number of blacks had surpassed that of the foreign born in Chicago and Philadelphia. The capital city was more than half black, Baltimore over one-third, and Philadelphia and Chicago approximately one-fourth black.[2]

Blacks and foreign immigrants who pressed into the cities, particularly the inner cities, had much in common: they held the lowest-paying jobs, they were the last hired in good times and first fired in bad, and they were often crowded into neighborhoods of dilapidated houses and poor sanitation. Poverty, filth, and ignorance combined to breed disease. The migrants were thus prime candidates for the wards of city hospitals.

In response to mounting needs, local governments increased the number of hospitals under their control from 915 in 1923, the first year for which comprehensive figures are available, to 1,324 in 1960. New hospitals and enlarged older ones swelled the number of beds in these institutions from nearly 116,000 to over two hundred thousand, an increase of 74 percent in the number of beds provided by local governments. (During the same period the number of beds in non-government-owned hospitals increased by 86 percent—from 283,000 in 1923 to 527,000 in 1960.) Some city hospitals grew enormously. In 1950 the five largest ranged from Kings County in Brooklyn, with 2,509 beds, to Cook County, with 3,460. Los Angeles County Hospital admitted over seventy-seven thousand patients per year and Cook County Hospital over eighty thousand.[3] How the city hospitals met the challenges of an era of increasing needs, burgeoning medical knowledge, and multiplying techniques is illustrated by the histories of several of the larger city hospitals.

The Boston City Hospital entered the second decade of the twentieth century with a strong medical staff backed by two medical schools, a nationally recognized school of nursing, and a substantial physical plant that accommodated over one thousand patients. In 1910 nearly fifteen thousand patients were treated on its wards at a cost of over $500,000 (see Table 7). Morale was high. A leading American journal in the hospital profession called the institution "the peer of any in the country in architecture, in equipment, and in administration." In the late nineteenth century the board of trustees, the staff, and the administration, aided by social and political connections, had managed to obtain reasonably adequate funds for the hospital. But the old order was changing, in Boston as elsewhere. As the city had grown, political control had become more widely diffused and the machinery of government had become unwieldy and thus unresponsive.[4]

In an attempt to improve the efficiency of Boston's government, the various administrators and boards were placed directly under the mayor in 1885, and in 1909 the right of the city council to interfere in his administration was drastically curtailed. Such centralization of power was part of a national trend that began in Brooklyn, New York, in 1882 and eventually included the rest of New York City and Philadelphia, among other cities. This opened the way for a single man to control the machinery of government, and the Irish immigrants who had arrived in Boston by midcentury conferred that power upon a succession of political bosses. The most noted of them was James Michael Curley, who dominated the politics of the city from 1913 to 1949, becoming one of the most powerful of America's political monarchs. From the start the so-called "mayor of the poor" began a program of spending for public works that was to continue throughout his political career.[5]

Curley's special favorite was the city hospital, which stood near his boyhood home. Because, he claimed, he had known what sickness, suffering, and hunger were, he made sure that Boston City became "one of the world's half dozen great hospitals."[6] He encouraged administrators and staff to plan new buildings and to ask for modern equipment, and he sold these improvements to the politicians and the voters. Curley also appointed a capable health commissioner for the city of Boston and supported him so well that Boston headed the list of American cities graded for their performance in public health by the American Public Health Association.[7] He supported the professional work of the hospital as well. The story is told that at the beginning of Curley's reign as mayor, Frank B. Mallory's yearly budget to run the pathology laboratory was a mere $500. Curley sent for Mallory, asked a number of questions, and was impressed by the pathologist and his work. From that time on, appropriations for

**Table 7. Selected statistics on four public general hospitals, 1910–1979.**

| Year | Capacity (beds) | Admissions | Average daily patient census | Births | Total expense (thousands of dollars) |
|------|------|------|------|------|------|
| Bay View Asylum—Baltimore City Hospitals[a] | | | | | |
| 1920 | 1,800 | — | 1,162 | — | 412* |
| 1930 | 1,052 | 4,456* | 905 | — | 479* |
| 1940 | 1,379 | 8,269 | 1,102 | 2,293 | 764* |
| 1950 | 1,810 | 7,251 | 1,395 | 2,612 | 2,258 |
| 1960 | 2,063 | 11,408 | 1,276 | 4,854 | 5,986 |
| 1970 | 857 | 11,792 | 612 | 3,022 | 18,403 |
| 1979 | 374 | 10,425 | 261 | 1,283 | 37,385 |
| Boston City Hospital[b] | | | | | |
| 1910 | 1,061* | 14,442* | 725* | — | 536* |
| 1920 | 1,202 | 19,383* | 763 | — | 1,239* |
| 1930 | 1,593 | 32,476* | 1,139 | — | 2,898* |
| 1940 | 2,392 | 43,620 | 1,389 | 3,177 | 3,284* |
| 1950 | 2,376 | 36,307 | 1,856 | 2,991 | 9,185* |
| 1960 | 1,325 | 29,375 | 944 | 3,234 | 16,154 |
| 1970 | 859 | 23,789 | 658 | 2,355 | 41,412 |
| 1977 | 454 | 16,206 | 354 | 1,679 | 64,230 |
| Cook County Hospital[c] | | | | | |
| 1910 | 1,350* | 28,932* | 1,200* | — | 411* |
| 1920 | 2,700 | 32,288* | 1,830 | — | 1,601* |
| 1930 | 3,300 | 51,414*[d] | 2,339 | — | — |
| 1940 | 3,300 | 81,154 | 3,108 | 4,818 | 4,540* |
| 1950 | 3,260 | 80,207 | 2,828 | 7,367 | 4,783 |
| 1960 | 3,200 | 85,126 | 2,422 | 18,662 | 11,023* |
| 1970 | 2,263 | 64,171 | 1,628 | 11,452 | 54,511 |
| 1979 | 1,363 | 53,028 | 964 | 7,017 | 123,856 |
| Washington Asylum Hospital—Gallinger Municipal Hospital—District of Columbia General Hospital | | | | | |
| 1910 | 175*[e] | 2,575*[e] | 124* | — | 70*[e] |
| 1920 | 260 | 2,587* | 169 | — | 118* |
| 1930 | 560 | 7,454* | 405 | — | 594* |
| 1940 | 1,392 | 15,881 | 903 | 1,929 | 871* |
| 1950 | 1,416 | 19,521 | 1,002 | 4,871 | 4,490 |
| 1960 | 1,121 | 23,868 | 919 | 6,519 | 10,984 |
| 1970 | 983 | 18,675 | 602 | 5,094 | 24,373 |
| 1979 | 492 | 9,732 | 416 | 1,297 | 47,497 |

Sources: All figures except those marked * are from lists published by the American Medical Association (Council on Medical Education and Hospitals, "Hospital Service in the United States") in the *Journal of the American Medical Association* for 1920–1940 and by the American Hospital Association (*Hospitals,* Guide Issue for 1950–1970, and *Guide to the Health Care Field* for 1979). Those marked *, unavailable in these lists, are from the annual reports of the individual hospitals, with the exception of the 1910 figures for Cook County Hospital, which are from Henry Burdett, *Burdett's Hospitals and Charities* (London: Scientific Press, 1913), p. 767.

a. Includes psychiatric, tuberculosis, and chronic disease hospitals; probably also includes the infirmary in 1920.

b. Includes all departments except sanatorium division.

c. Includes psychiatric hospital except in 1930.

d. This figure is for 1931; annual report for 1930 is unavailable.

e. These figures are for fiscal year 1913–1914; annual report for 1910 is unavailable.

pathology were most generous. Curley stressed his personal involvement by having members of his family cared for in the hospital and by parading around the institution like a general inspecting his crack regiment. He sent letters to house officers praising them for the care they gave to particular patients, and when his own end came, it was after an operation in his "personal" hospital.[8]

Through it all he remained a politician: he expected favors in return. His henchmen and their families were cared for without charge in the city hospital. He would take a basket of food from the hospital kitchen to a family in need and assume credit for the gift. Nonprofessional employees such as orderlies, charwomen, and laborers held patronage jobs, and during Curley's regime their jobs were made sinecures by the simple expedient of appointing many more than were needed. Thus, they had plenty of time for political errands as well as for loafing, fraternizing, and just plain talking.[9]

An entry from my diary, made in 1934, when I was on the staff of the Boston City Hospital, reads:

> Without doubt, one-third as many orderlies, porters, messengers, charwomen and kitchen-men could do the amount of work that is done by the present number, and do it twice as well—if they worked steadily and kept their minds on their jobs during their hours of employment.
>
> During the winter months it is necessary for all of us to travel from one building to another through the basement corridors; and here, all day long, one may find any number of these helpers, talking, laughing, shouting, in groups of two or three, usually. When one of them happens to have a particular task in process, it makes little difference if there is an opportunity to do something else. Hence one frequently sees an orderly stop pushing a litter containing a patient, in order to have a little friendly gossip. Often one hears a chorus of raucous laughs, as the joke ends—the joke which has been passed over the head of some feeble human being, sick unto death.
>
> The helpers come out of the kitchen, two or three at a time, to loaf in the corridor, and rub their dirty aprons against its walls, before they go back to handle the food again. The charwomen always manage to manipulate their scrubbing so that two or three can mop continually in the same neighborhood. Then it is an easy task for them to gather together for a little friendly chat. Thus we see them, most of the time during the day, it seems, with their heads bent close together and their mops reaching out from the center to the floor in all directions—for all the world like a gathering of witches over a tripod.
>
> The most amazing quality of these people is their imperturbability in the face of anything or anybody. It is not unusual for a group of doctors to board an elevator, after the usual interminable wait, and find that they are after all only of minor importance, because it is necessary for the elevator operator to continue her conversation with her "girl friend" all the way up to their destination (if she remembers which floor it is).
>
> These people seem to know, instinctively, that their positions were obtained by political power and can be held only by the same methods. The amount of work done is an extremely unimportant consideration....And

the people higher up? Do they know? Well, they know that they can command a certain number of votes. Why should they care about anything else?

Meanwhile, the people pay.[10]

A $3-million building project, begun in 1924, produced new buildings for obstetrical, gynecological, and surgical patients, new residences for nurses and house officers, and the enlargement of other ward buildings and service buildings. By 1930 Boston City Hospital had facilities for nearly 1,600 patients (see Table 7), and one-tenth of Boston's childbirths took place within its walls, more than in any other hospital in the city.[11] Another $1.5 million committed to construction in the latter year produced a new pediatrics building, named after Curley's wife, a magnificent institute of pathology, and renovations in other areas. Practically the entire hospital had been rebuilt in twenty-five years, and most of the improvements had occurred while Curley was in office. Since his successors recognized the vote-getting power of a popular city hospital, its physical growth continued for several more decades, and the operating budget accelerated with the growth. By 1955 the hospital's budget was nearly $16 million a year for an average daily census of 2,447 patients, compared with Cook County Hospital's annual expenditure of less than half that for a slightly larger number of patients ($7.8 million for an average daily patient census of 2,529).[12]

Despite all the improvements, demand often exceeded capabilities, equipment continued to wear out and become obsolete, buildings deteriorated, and space was inadequate for the ever-increasing numbers of patients and personnel. Furthermore, Boston's finances were feeling the pinch. In the late 1950s, when the hospital was offering some of the finest medical care in the country and its educational and research programs were winning worldwide fame (see Chapter 6), some people were advocating its closing because the increasing costs of medical care were putting too much strain on the city's funds, which were suffering from a decreasing tax base as more and more people moved to the suburbs. In desperation the city resorted to a reorganization of its entire health-care system to solve the dilemma (see Chapter 8). The new system, which began in January 1966, was expected to coordinate ambulatory and hospital care and make services more accessible, but whether it would ease the strain on the city's budget remained to be seen.[13]

While Boston City Hospital had been enjoying its golden age, the other large city hospital that opened after the Civil War was struggling to retain its reputation as a satisfactory provider of medical care. Cook County Hospital admitted 28,932 patients during 1910, straining to the limit its capacity of 1,350 beds (see Table 7). The assistant superintendent of Johns Hopkins Hospital called it the worst municipal hospital in the United States, considering the resources of the city in which it was situated.[14] The next half century saw the hospital continually striving to make enough room for the throngs who came to its doors, but never succeeding. An impressive number of buildings were erected: in 1914 a psychiatric hospital; in 1916 a main building a block long and eight stories high, which contained administrative offices, operating rooms, and many large wards; and by 1928 a four-hundred-bed children's hospital, claimed to be the largest medical institution for children in the world, a men's pavilion, and a building for clinical and pathological laboratories.[15]

But imposing piles of bricks do not a hospital make. A look inside told a different story. In 1927 surveyors from the American College of Surgeons found the wards to be dirty, crowded, in ill repair, short of essential supplies such as bed linens, and lacking in standard equipment. The wards were poorly planned and too large for efficient and humane medical care, and the aisles were choked with extra beds. The small corps of graduate and student nurses, though highly efficient, was unable to cope with the sheer volume of work.[16] This report was followed by denials, charges and countercharges, more surveys, and little action. In 1935, 1936, and 1937 the American College of Surgeons granted the hospital only provisional approval and in 1938 withdrew approval altogether, a disgrace for city and hospital, a tragedy for nurses and doctors training there and a calamity for patients. (See below for the origin of the accreditation program for hospitals.) In 1940, after a little patching and painting, a few additions to housekeeping and nursing staffs, and the establishment of an outpatient department, approval was granted again, but shortages, problems of housekeeping, and poor management still continued. Meanwhile, construction of buildings went on: a nurses' school and residence opened in 1934 or 1935, a power plant and laundry in 1939. Adjacent buildings were purchased for a larger outpatient department and for a research institute.[17]

By 1930 the number of beds in the hospital had more than doubled, reaching 3,300; by 1960 the capacity was essentially unchanged, but the number of admissions had nearly quadrupled to over eighty-five thousand a year, including nearly nineteen thousand births (see Table 7). Wards were still dirty, ill-kept, and crowded; nurses were still too rushed to give much more than a lick and a promise; and house officers were so busy putting out brush fires that they had little time for the careful study needed for very sick patients, of whom they had many. While the advancing science of medicine was calling for more and improved facilities and was promising greater rewards than ever before, Chicago's city hospital was falling farther and farther behind.

Why did this huge hospital in one of the world's richest cities fail to provide a decent home for the underprivileged people condemned to go there? We can only conclude that the Board of Commissioners of Cook County wanted it that way. Instead of placing the hospital under a specific governmental unit or a separate board of citizens charged with the responsibility for running it, the commissioners had retained direct control for themselves. Thus, they could dispose of funds available from patients and insurance companies, from the state and federal governments, and from bond issues that they persuaded the public to vote for periodically. They appointed subprofessional workers (who numbered nearly 2,200 in 1960) by the patronage system.[18] And, significantly, they could award contracts for renovations and new buildings and for mountains of supplies and equipment each year. There is no reason to believe that opportunities for personal profit from such transactions had disappeared following the exposure of the boodle ring in the 1880s.[19]

Even though they assumed the duties of a board of trustees for this huge hospital, the commissioners displayed a cavalier attitude toward the institution much of the time. Surveyors of the American College of Surgeons found in 1932 that the hospital committee of the board of commissioners did not meet at prearranged intervals as hospital governing boards customarily do but met when it was so inclined. Nor did the hospital or the commissioners publish detailed financial reports covering the activities of the hospital. The eighty-two page annual message of the president of the Board of Commissioners of Cook County for fiscal 1932 contained only a half page of very general figures on the finances of Cook County Hospital. The trustees of the Boston City Hospital, in comparison, published sixteen pages of financial details and elaborate tables in their twenty-four-page annual report for the same year. Just as clinical records are a good index of the quality of medical care, so a thorough, detailed annual report is indicative of a well-run hospital. As late as 1970 Pierre DeVise of Metropolitan Chicago's Health Planning Council could still contend, "No one knows how much money is spent by Cook County Hospital, how much it costs to serve patients, how many people seek care at the hospital, or how many beds are in the hospital."[20]

To keep this vast enterprise operating so that the county commissioners were free to pursue other interests, there had to be at the helm someone who was knowledgeable, an expert manager, someone who commanded sufficient respect or obedience from the various groups—doctors, nurses, politicians—to keep them working together and to keep the protests of newspapers, citizens' groups, and medical societies to a minimum. The man who accomplished this task for over half a century was Karl Meyer—surgeon, administrator, politician, and an expert in each field. Meyer began his association with the hospital as an intern in 1908, and it is characteristic of him that he wrote first in the examination for internship. Around this time the hospital was being criticized in both the medical and the lay press because house officers were not adequately supervised, so in 1914 Meyer was appointed medical superintendent to remedy the situation (see also Chapter 3). As he tackled the job of fitting together a set of idealistic rules and a corps of overworked and independent house officers, he developed into a shrewd and tough-minded yet flexible and pragmatic administrator. At the same time he refined his surgical skills and kept them sharpened by operating at Cook County Hospital, where he had the pick of interesting and complicated cases, and by becoming chief of surgery at a private hospital. It seemed impossible to supervise adequately the work of all the house officers, act as chief of a surgical ward, and carry on a private practice, but Meyer managed to do all of these sufficiently well to keep criticism of the hospital below the eruption point. Over the years he gradually gathered the reins of power so that it was he who was making the policy decisions and not the warden, who had traditionally been the chief executive officer.[21]

Yet, those who probed found much to censure. In 1936 a Chicago newspaper labeled the hospital "Misery Harbor" and described graphically the dirt, vermin, and poor food, the scanty supplies, ill-kept records, and broken equipment. Meyer and the hospital were able to weather these and other criticisms. He demonstrated his political endurance especially in the 1930s, when the administration of the hospital was challenged from several sides. A committee of the American Medical Association denounced the "lamentable absence of a broad, comprehensive, coordinated public hospitalization program" for the Chicago area. A committee of Chicago doctors contended that the staff structure of the hospital resulted in inferior medical care and recommended its replacement with university-type services (see Chapter 6 for a discussion of university services). Finally, the American Medical Association and the American College of Surgeons both removed Cook County Hospital from their approved lists. The reaction of the county commissioners was not to dismiss the man who was actually responsible for affairs at the hospital; instead, in December 1939, they created a new position, medical superintendent of Cook County institutions, and elevated Meyer to it. From this broader power base he continued to run Cook County Hospital as before, retaining his office and private apartment in the hospital and his position as chief of a surgical service.[22]

A citizens' committee investigating the hospital in 1937 complained that no complete table of organization existed. They implied that, just as the warden's title was more suitable for a jail than a hospital, so also he now needed experience in administering a medical facility; present and previous wardens had been untrained in hospital administration. But the committee either did not recognize or did not care to expose the real situation: that the warden merely took care of the physical plant and housekeeping and in other matters followed the orders of Karl Meyer, the actual administrator of the hospital. The committee did suggest, however, that a director with broader authority than the warden be appointed and that he be given several well-qualified assistants.[23]

Yet, as late as 1964 so little change had been made that another citizens' committee, retracing this much-traveled ground, called supervision at the hospital weak from top to bottom and charged that administration of the hospital lacked unity and consistency. They recommended that a professionally qualified hospital administrator be placed in complete administrative control. In the following year this was finally done.[24]

Meanwhile, those who ran the hospital, without setting their house in order continued to build the house bigger and bigger—not only hospital buildings, but a thirteen-story residence hall for house officers, which was said to surpass "in comfort any dormitory for interns and residents anywhere in the world." They built showcase units, such as a research institute and a postgraduate school of medicine, and established model burn and trauma units.[25] But they skimped on supplies and equipment, appointed slack workers at the subprofessional level through the patronage system, and exploited nurses and house officers while boasting to the public that a visit to the hospital would impress "even the most casual observer that here is an institution of which to be proud." In politics, as Theodore H. White has stated, "it is not the way things really are that counts, but the way they appear to be." Affairs at Cook County Hospital had appeared to be all right to enough people as long as Karl Meyer was active; when he retired in 1967, the institution, which had been patched together for so long, began to fall apart.[26] By 1970 many Chicagoans looked at the hospital as "an unsanitary dumping ground which barely manages to get accredited every three years." Its physical plant was called "a hopelessly obsolete labyrinth of . . . unwashed tunnels . . . a nightmarish caricature of a community hospital 10 times too large," where patients waited an average of two hours to be examined by a doctor.[27]

The ingenuity that kept this hospital going until the 1960s generates a certain awe, but one cannot help deploring the tragedy that the system brought to so many patients and the time and energy wasted by doctors and nurses who tried to work amid its inefficiencies and frustrations. Yet, by observing Cook County Hospital in the twentieth century, we can understand better what patients and staffs of city hospitals had to endure in the nineteenth and the eighteenth centuries.

Boston City and Cook County Hospitals are examples of city hospitals that were manipulated to support and maintain the power of political groups—with different results. But while they had their problems, they did not bear the added burden of an almshouse. By contrast, in the early twentieth century the city hospitals of Washington and Baltimore were still freeing themselves from the vestiges of almshouse influence, and they did not conform to the pattern of the typical city hospital until the 1930s. In fiscal year 1913–14 the Washington Asylum had an average daily census of 124 patients (see Table 7) and consisted of eleven buildings scattered along a hillside, plus the jail and homes for the superintendent and his assistant, who were still in charge of both hospital and jail. The general hospital consisted of four frame buildings of Civil War vintage, each a single, open ward intended for thirty patients but often packed with as many as fifty. They were the same buildings that had been labeled "a disgrace to the capital city" by hospital consultants in 1897. Patients had to be wheeled back and forth through snow, rain, and cold to a separate, barnlike building for surgical operations. This structure also contained the maternity unit, a ward for chronically ill patients, and the tiny hospital laboratory. Separate buildings housed the receiving ward and psychopathic, smallpox, and other isolation patients.[28]

Still governed by the five-member Board of Charities, the hospital was always dependent on the whims of Congress for its funds. Although appointed by the President of the United States and responsible to the commissioners of the District of Columbia,[29] the board had to conform also to the wishes of several congressional committees and contend with their indifference, which was exemplified by the remarks of Senator Harlan J. Bushfield of South Dakota while investigating Gallinger Municipal Hospital in 1943. He informed a witness that none of the members of the subcommittee on the hospital "resides in the District of Columbia; we have no particular interest in Gallinger Hospital, and as a matter of fact not a one of the three of us had ever heard of it before we were appointed to this committee."[30] At any time the President's staff or a congressional committee might intervene in an action of the commissioners or the hospital board. An example was the appointment of James A. Gannon as chief physician at the Washington Asylum Hospital.

On March 31, 1914, President Woodrow Wilson's secretary, Joseph P. Tumulty, telephoned Oliver P. Newman, chairman of the Board of Commissioners of the District of Columbia, to ask why a candidate recommended by the White House for the position of visiting physician at the Washington Asylum Hospital had not been appointed. According to Gannon:

> Mr. Newman replied that among the other names being considered several had been endorsed by members of the District Committees of the House and the Senate and that the Board of Commissioners were still considering the matter. Mr. Newman further stated that they would not name the

White House candidate unless the President particularly desired that they do so. Mr. Tumulty answered: "The President particularly desires that they do so."

On the following day Gannon received the appointment.[31]

While sharp eyes in congressional committees and in federal bureaus helped prevent major graft such as occurred in other large cities, Washington's city hospital suffered instead from the preoccupation of the executive branch with national and international affairs. When congressmen thought of the hospital at all, they tended to view it as a place to cut appropriations without losing votes. In addition, southern congressmen resented that money was being spent on blacks, who accounted for 49 percent of the admissions to Gallinger Hospital in 1915 and 57 percent in 1935.[32] Southern congressmen controlled District of Columbia appropriations because of seniority. Too often their attitude was like that of the chairman of the House Subcommittee on District Appropriations, Ross Collins of Mississippi. In 1939, when the head of the District of Columbia welfare department sought assistance from him, the congressman, looking the social worker straight in the eye, said, "If I went along with your ideas, Mr. Street, I'd never keep my seat in Congress. My constituents wouldn't stand for spending all that money on niggers."[33]

Despite repeated rejections, the Board of Charities persisted in asking for funds for a general hospital. Year after year they sent requests to Congress, and year after year they were passed over. Meanwhile other hospitals in the District of Columbia were reaching members of Congress. During the first six years of the twentieth century Congress gave over $500,000 to private hospitals in Washington for new buildings and reconstruction of old ones. The Board of Charities protested that although this money had "somewhat improved the facilities enjoyed by indigent patients in these hospitals," by far the greater part of it had provided "administration buildings, magnificent operating rooms, and luxurious accommodations for private patients able to pay."[34] Instead this sum could have been used to erect a modern hospital run by the District of Columbia government.[35]

Finally, by changing the hospital's name to Gallinger Municipal Hospital (both to minimize the poorhouse image and to curry favor with Congress by naming the institution after Senator Jacob H. Gallinger, who had been chairman of the Senate District Committee for many years), the city authorities succeeded in getting funds for a general hospital. But a series of setbacks prevented early construction. First came vigorous protests and demonstrations by citizens in the northwestern section of the city, who feared a loss of property values near the site projected for the hospital in that area. Then construction

was delayed by the First World War. At long last, in 1929 a general
hospital of nearly three hundred beds was opened at the asylum site.
During the next two decades buildings were constructed for women,
children, patients with tuberculosis and contagious diseases, and
outpatients. In 1930 the hospital could accommodate 560 patients and
in 1950 1,416. To run this modern institution, trained hospital
administrators were appointed beginning in 1927.[36]

Patients flocked to the rejuvenated hospital. The average daily
census climbed from 405 in 1930 to 903 in 1940 (see Table 7), an
increase of over 120 percent in a decade. Neither space, staff, nor
budget could keep up with such growth. The authorities did their
best: no sooner were patients moved into a new building than the
ward they left was filled with others. Yet, the hospital, though running
hard, could not catch up. Despite newspaper articles throughout the
1930s complaining of overcrowding, outmoded facilities, and shortages of nurses and other help, it took a congressional investigation in
1943 to demonstrate that the budget was too low for service units
such as kitchen and laundry, that the X-ray department was too small
and too old, that space for patients was still lacking in some areas, and
that the number of positions for doctors, nurses, and attendants was
inadequate.[37]

In the 1950s an outspoken chief of staff, Joseph Fazekas, and
protests by civic-minded house officers managed to force higher
budget requests than city officials had originally proposed. By the
early 1960s it was clear than an extensive program of new building and
modernization was required. To prepare for the new buildings a 105-
year-old structure was demolished. It had successively housed Civil
War wounded, insane patients, vagrants, crippled children, adult
patients, nurses, classrooms, and finally house officers.[38] Did its
removal symbolize the end of making do and the beginning of a new
era of adequate facilities? Patients—and all those who cared for
them—hoped so.

Another hospital that changed radically in the twentieth century
was Bay View Asylum in Baltimore. Throughout the previous century
it had remained mostly an almshouse and poor farm with psychiatric
and chronic hospitals attached. The small but increasing number of
acutely ill deposited on its doorsteps as the city had grown had been
stowed away in poorly arranged wards in the chronic hospital. But
change was in the making: in 1895 a victory by reform forces over
Baltimore's long-dominant political machine opened the way for a
new city charter in 1898. A nine-man Board of Supervisors of City
Charities was created to replace the Trustees of the Poor, and Bay
View was placed under the new board. A wave of social, sanitary, and
health reforms followed, during which the city built three hospitals:
Sydenham, for contagious diseases, in 1909, and at Bay View a

tuberculosis hospital in 1904 and a general hospital for acute diseases in 1911. Medical and surgical services were organized in conjunction with the two Baltimore medical schools, Johns Hopkins and the University of Maryland, and training of nurses was begun (see Chapters 4 and 6).[39]

After a few years, however, the forward movement faltered. The board of supervisors fell into the hands of "minor politicians chiefly concerned with peanut jobs which had some relation to the purchase of supplies." In 1921 J. Whitridge Williams of Johns Hopkins Medical School resigned from the board because jobs were being awarded for political reasons. Two years later brawling over the division of spoils made so much noise that the public demanded and got a more committed board of supervisors headed by a distinguished banker. Even more significant for the welfare of the general hospital was the appointment to the board of George Walker. A urologist well known in social circles, he was immediately made chairman of the Bay View Hospital Committee. Honest, courageous, and loyal to his city, he believed in taking direct action to right an unjust wrong, even to the extent of threatening to kill the petty, inept, and arbitrary command-ing officer of the Johns Hopkins hospital unit in France because he had practically ruined morale. The problems he confronted at Bay View were equally grave. The hospital was dirty, vermin ridden, and in need of repairs. Crowding was episodic in the acute and continuous in the chronic and psychiatric hospitals. The food was the poorest to be found in several city hospitals investigated; in 1922 a board member saw an evening meal served that consisted of black coffee without sugar and two slices of dry bread. Apparently, proximity to the almshouse had brought poorhouse conditions to the hospital; the hogs on the poor farm seemed to get more food than the patients. Since low salaries for nurses and orderlies kept many jobs unfilled, the hospital depended in large part on inmate help not only for cleaning and labor but for the care of the sick.[40]

In 1924 the supervisors replaced an inefficient superintendent with a retired army officer, Rufus E. Longan. He set to work to cleanse the Augean stables. Among the political appointees he fired was the son-in-law of the Democratic boss of the city, and the governing board backed him. He improved the physical plant, instilled a modicum of efficiency into the inmate labor, kept the antiquated wards as clean and orderly as possible, stretched the small budget as far as it could go, and in general ran a tight ship. Yet, although the institution's name was changed to Baltimore City Hospitals in 1925, the hospital remained the tail on the almshouse dog. The number of beds in the general hospital was only one-tenth of the sixteen hundred in the entire institution, and even after the hospital was in full operation, its patients were still referred to officially as inmates. The administration was more accustomed to managing dormitories and a poor farm than it was skilled in handling the special problems of a hospital, and the budget was geared more to maintaining the poor at a subsistence level than to treating sick persons at the only acceptable standard—the best. The amount spent per patient for one day's care in the general hospital was only $1.65 in 1925, much lower than at many other city hospitals. The same year the amount was $4.30 at Boston City, $3.60 at Gallinger, $3.09 at Bellevue, and $2.99 at Cook County (see Table 11 in Chapter 7.)[41]

When the Depression of the 1930s hit, the board of supervisors and the superintendent cut their already low budget to the bone. This had to suffice, for the leadership of "untidy, lethargic, hospitable, well-fed, contented, happy Baltimore" would not call upon their citizens to do more. Having built the new hospital, the city mostly ignored it, except when a victim of poverty or a psychotic patient had to be disposed of or an automobile accident happened nearby. The average citizen of Baltimore was barely aware of the hospital's existence.[42]

The hospital had several glaring deficiencies: it had no departments for gynecology and obstetrics or for children and no outpatient department. It lacked certain vital equipment: X-ray equipment was not installed until 1919, and even then it could not be used because there was no salary for a technician. Furthermore, as far as can be determined, there was no electrocardiograph until the new acute hospital building was opened in 1935. (In contrast to Bay View, Boston City Hospital was a pioneer in the use of X-rays, and Cook County Hospital purchased its first X-ray apparatus in 1902, while electrocardiographs were introduced at Boston City Hospital in 1920 and Cook County Hospital by 1923.)[43] Finally, there was no training school for graduate nurses, only a school for practical nurses. The general philosophy of the supervisors seemed to be to do well those things they chose to do and to leave other things alone. Such omissions and shortcomings could be tolerated only by dependence on other hospitals. In 1930, for instance, the city hospital was providing only 25 percent of the care to the indigent sick supplied by hospitals in Baltimore; the two university hospitals and private hospitals supplied the rest. Likewise, if an electrocardiogram was urgently needed and the patient could stand the trip, the Johns Hopkins Hospital might be requested to perform the test.[44] Dependence upon other hospitals for important laboratory tests was a makeshift arrangement that seldom produced satisfactory results, because only a fraction of the city hospital patients who needed tests received them.[45]

The governing board took care to avoid political entanglements, while the doctors considered politicians pariahs. Thomas R. Boggs, chief of medicine, once instructed his chief resident to tear up the internship application of an otherwise suitable medical student who had obtained a letter of endorsement from a city councilman. Consequently, politicians merely shifted their attention to more fertile fields, where they could sow the income from taxes and reap the profits. Toward the city hospital their attitude seemed to be: no favors, no money. The supervisors and the staff were left with righteousness, and near starvation. When Longan retired as superintendent in 1933, he told a reporter, "If I had known when I came here that I would have had so many problems, there are not enough dollars in the banks of Baltimore to attract me to the job."[46]

The only recourse for the board of supervisors was to enlist the aid of influential citizens to obtain funds for the city hospital. This drive was led by George Walker; largely through his efforts the state legislature authorized a bond issue for new buildings. A nurses' home was opened in 1931 and a new general hospital in 1935. Into the latter were moved the medical and surgical services, to which obstetrical and pediatric services were added, while in the former acute hospital were placed wards for chronically ill patients, removed at last from the almshouse.[47]

The new era was inaugurated in 1935 with the placement of the hospital under a recently created board of welfare, which set in motion the policies that the board of supervisors had tried to pursue since 1923. For the first time a trained hospital administrator, Parker J. McMillin, was employed. Under his administration the transformation into a comprehensive hospital for acutely ill patients was completed, improvements were made in the chronic hospital and infirmary, the caretaking functions of the psychiatric hospital were taken over by the state, while the tuberculosis service accelerated its activities to utilize the highly effective antituberculosis drugs introduced in the 1940s and 1950s. Old buildings were renovated and occupied, and in 1950 some tuberculosis patients were placed in Sydenham Hospital, which had been closed as a separate institution for infectious diseases and its patients transferred to the city hospital after the introduction of antibiotics. An outpatient department was added in 1950 (see Chapter 7).[48]

Construction in the next two decades included a new tuberculosis hospital, an addition to the acute hospital, and renovations of several buildings, including the large infirmary building, which dated from the opening of Bay View Asylum. The old open wards were converted into private and semiprivate rooms to conform to modern concepts of medical care and human dignity. These ward alterations began in the 1950s, before the Johns Hopkins Hospital began similar modifications. The metamorphosis of the city hospital enabled it to compete favorably with private hospitals as the advent of Social Security and the rapid growth of medical insurance in recent years greatly increased the number of paying patients who sought admission (see Chapter 7).[49] The facilities of the hospital were improved in other ways as well. The business services were modernized, and a centralized computer service was installed that made possible a rapid comparison with the activities of other hospitals in Maryland and throughout the United States. The hospital took advantage of its spacious grounds to build garden apartments for the house staff and thus was able to compete favorably for house officers.[50]

As the removal of psychiatric and extended-care patients to state institutions freed the hospital from the last traces of the almshouse, there was no longer any justification for keeping an agency for health care in a department of public welfare. In 1965 it was given separate status as the Department of Hospitals. In that year, in this institution of 1,729 beds, 14,644 patients were admitted and 3,464 babies delivered at a total cost of $10,778,000.[51] The city hospital of Baltimore had changed slowly at first, but in the half century following 1910 it had completed its transformation to a hospital rendering first-class medical care. In addition, it was beginning to heed the compelling challenge of recent years to serve the community more.

The examples of these four hospitals show how local governments related to the hospitals they controlled during a period of increasing demand for hospital services. In the twentieth century local governments became gradually more efficient, although they still had a long way to go. Cities placed administrative functions under a single person and eventually surrounded him with a staff of subexecutives. Superintendents of city hospitals whose only virtue was party loyalty were gradually replaced with experienced hospital administrators. George H.M. Rowe, who was at the Boston City Hospital from 1879 to 1907, was one of the earliest and most successful of these administrators. Comparable chief executives were not appointed in the city hospitals of Washington and Baltimore until 1927 and 1935. In fact, as late as 1921 the board of supervisors of Bay View rejected a motion that institutional training other than penal be an essential qualification in selecting a hospital superintendent and voted that a medical superintendent was not needed. They were still thinking in poorhouse terms even though they had opened a general hospital ten years before. Cook County Hospital was even slower to place a trained administrator in authority, not appointing an overall director until 1965.[52]

Above the superintendent or warden of the city hospital or almshouse of the eighteenth and nineteenth centuries there was often a board of citizens representing the public. As hospital administrators became more knowledgeable, medical staffs more highly organized, and doctors more skillful in diagnosis and prognosis, these groups made the decisions regarding admission and discharge and regulated the activities of staff and employees, while governing boards concentrated on policy decisions. In 1897–98 the trustees of the Boston City Hospital were formally prohibited from admitting patients or dismissing them once admitted.[53] When the administration of public welfare became more highly organized, city hospitals connected with almshouses were often placed under municipal boards of charities or welfare, and sometimes charities and correction were combined under one board (see Table 3 and Chapter 2).

As hospital administrators grew in expertise and prestige, they were given more autonomy, whether within city or county departments or under boards of citizens. When the administrator was answerable to a department of the local government, the hospital or department was usually aided by an advisory board of citizens. If the hospital or welfare board was fortunate, the members of these governing or advisory boards acted as links with influential and powerful members of the community. In Baltimore, for instance, a widespread network joined business, social, and religious leaders, the educational establishment—especially Johns Hopkins University—and those interested in health and welfare. This informal linkage was reasonably effective when Baltimore was a small city and power resided in a few hands. After the hospital was placed under the Department of Public Welfare in 1935, the director of the department relied heavily upon a medical advisory board in matters relating to the hospital. This board included members of the welfare department, the health commissioner of the city, the deans of the medical schools, and, from the hospital itself, the superintendent and chiefs of the medical services. The new organization was advantageous in that it placed control in the hands of professionals in health and welfare fields. Yet, participation by civic leaders was diminished, and thus the hospital became more insulated from the public.[54]

In Boston the membership of the city hospital's board of trustees shifted from Boston's Brahmins to first- and second-generation immigrants, particularly the Irish, as the power structure of the city changed hands.[55] Thus, in the early twentieth century the board represented the dominant political groups in the city, although it became increasingly separated from those who were in a position to give large sums of money. As late as 1959 an astute chairman, recognizing the need for representation of Irish and southern European groups, persuaded Richard Cardinal Cushing to join the board. The cardinal found this an effective way to serve his many parishioners who became patients in the hospital. He strongly supported programs for treating alcoholism, for instance, because of its frequency among the Irish.[56]

In Washington the Board of Public Welfare, which controlled the city hospital until 1937, and the advisory board, which was appointed when the hospital was placed under the health department (see Table

3), were kept busy trying to persuade members of Congress to meet the hospital's needs. Unfortunately, this left little time to keep the citizens informed. Partly for this reason and partly because of the multiplicity of interests in the nation's capital, the leading citizens of the city knew little and seemed to care less about the hospital.[57]

The group in control of Cook County Hospital kept the doors closed to business and society leaders. The few who were given any ingress were admitted only to the anterooms, such as the boards of the independent school of nursing and the research institute, not to the inner sanctum where the levers of power were manipulated.[58] To an outsider it appeared that hospital authorities kept the board of the school of nursing on a short leash, permitting its members little say in the affairs of the hospital itself, while only a minority of the governing board of the research institute represented the public.[59]

Political machines did not disappear during the era of the reform movements around the turn of the century. They lost some of their power as the civil service system began to displace patronage appointees and as professional training became a requirement for some administrative and technical positions in the health fields. But powerful machines continued to dominate big cities from time to time and thus to affect the fate of city hospitals. Curley's influence on the Boston City Hospital is just one example. Perhaps the most flagrant trespass upon the operations of a government hospital occurred at Charity Hospital in New Orleans when Huey Long controlled Louisiana politics, from 1928 to 1935. In 1913 the hospital had been placed under an independent board of administrators, which rescued it from political domination. The forces of good government maintained the board's independence until 1930, when Long, who had recently become governor, gained control of the board. Then, the hospital "reverted to the former vicious system." Long moved swiftly to replace the superintendent of the hospital with Arthur Vidrine, a young surgeon without administrative experience who had stumped for Long in his home parish. Later that year Long ordered the boards of Louisiana State University and Charity Hospital to create a medical school, a feat they accomplished in two hours.[60] Suddenly, drastic changes began to occur in the hospital's staff. Tulane University's program was severely curtailed, while Long pushed through a building program for the rival school. Charity became Huey Long's hospital as Louisiana State had become Huey Long's University. Only an assassin's bullet stopped this farce, and some may have seen retributive justice in the tragedy, for Arthur Vidrine was the surgeon who operated on Long and who failed to discover that Long's fatal hemorrhage came from a kidney, a diagnosis that a more experienced surgeon ought to have made.[61]

By contrast, Richard Daley, when he was political boss of Chicago and Cook County, exhibited a complete lack of interest in Cook County Hospital. When the Chicago police department came under fire in 1955, Daley searched the country for an outstanding chief and backed his expensive reforms. But the county hospital, which had subsisted on a starvation diet for decades and whose plight was plainly signaled every time it was inspected by an accreditation committee, was ignored by Daley, who plainly bore responsibility for the hospital as political leader of Cook County as well as mayor of Chicago. Obviously the Daley pyrotechnics would have been wasted on crumbling blockhouses that many voters perceived as receptacles for drunks, drug addicts, drifters, and derelicts. He and his henchmen correctly surmised that not enough people cared, and they acted accordingly.[62]

The story of the average city hospital tended to resemble that of Cook County Hospital much of the time, though there might be short periods of increased public interest, especially after a burst of newspaper publicity or the failure to gain reaccreditation brought civic leaders temporarily to the rescue. Life on such a seesaw is especially well documented for the city hospital in Washington, which was subjected to at least seven investigations between 1929 and 1960.[63] Boston City Hospital, on the other hand, benefited from its wide recognition in Boston and in international medical circles and was more plentifully suppled with funds than most city hospitals. In Baltimore the city hospital started the twentieth century with practically no attention, no friends, and no funds but gradually gained in prestige, support, and achievements.

Valiantly as the governing boards, administrators, and professional staffs of city hospitals fought to improve their standards, their greatest boost often came not from their friends in the city but from professional groups outside, especially state and national accrediting agencies. During the First World War the American College of Surgeons developed a program for evaluating the performance of hospitals and in 1917 formulated a set of minimum requirements for hospitals. These included adequate clinical and laboratory facilities to insure correct diagnoses, a thorough study and diagnosis of each case in writing, and a monthly audit of the medical and surgical work of the hospital.[64] This program of evaluation succeeded because enough people throughout the nation recognized that the potential consequences for a patient sick enough to be hospitalized were too grave to be trusted solely to the discretion and judgment of the individual doctor, as they had been in the past.[65]

The results of the first survey in 1918 were devastating: 692 hospitals of one-hundred-bed capacity or more were surveyed; only eighty-nine, or 13 percent, met the limited requirements of the American College of Surgeons. To avoid embarrassment, "the secretary was instructed to take the reports of the other 87 percent to the basement and burn them. The board feared the consequences of letting the true conditions in hospitals be known."[66] Obviously, city hospitals were not the only ones that needed to raise their standards. Within two years the situation improved enormously: 407, or 58 percent, of the 697 hospitals surveyed met the College's standards, and it became a primary objective of others to make the list, or become accredited, as approval came to be designated. Hospitals with certain deficiencies were approved provisionally and given a certain amount of time to correct them; 43 percent of all the hospitals approved were in this category. Of the approximately 160 city hospitals of one hundred beds or more, forty-five were approved, and of these seventeen, or 38 percent, were approved provisionally.[67] Six of the seven hospitals that have been discussed in detail were fully approved in 1920: the city hospitals of Baltimore, Boston, Chicago, and Philadelphia, as well as Bellevue in New York and Charity of New Orleans. Gallinger was not, for at that time construction of the first of its modern buildings had not even been started.[68]

Lack of approval or provisional approval served as a stimulus to staff, administration, and friends of a hospital to put pressure directly and through the news media upon politicians and the public to change procedures or grant the funds needed to gain approval. Friends of the hospital not only appealed to civic pride but also pointed out that an unapproved hospital would find it difficult or impossible to recruit satisfactory house officers and nurses, thus jeopardizing patients' lives. A good example was Cook County Hospital, which lost its accreditation completely for two years beginning in 1938 and was given only provisional accreditation from 1961 to 1963 and in 1968, 1970, and 1971.[69] Each time authorities bestirred themselves to correct the hospital's worst faults. Unfortunately, they never seemed to learn the full lesson: that wriggling into approved status should be only the beginning of full-scale reform.

The years from 1910 to 1960 were marked by great physical growth in city hospitals and increasing professionalization of hospital administration and a more definite place for these hospitals within city governments. The period was also characterized by more or less successful attempts by these hospitals to shake free from political domination and from the status of charity hospitals. These changes paved the way for improvements in medical and nursing staffs and advances in medical education and patient care. Yet, as patient demand increased, and as medical science, while promising more, also became expensive, there was never enough money.

# 6 · The Development of University Services

The teaching hospital is like a fish bowl;
the care of patients is under constant, critical review and scrutiny,...both its scientific and humane aspects...I know of no
better way to...assure the best care for
the citizens of a community, and thus the
best return for the health-care tax dollar,
than to provide...medical school-affiliated
hospitals.

Maxwell Finland

Where research is linked to practice, sick
people get better care.

Edward M. Dempsey

BY THE EARLY decades of the twentieth century
some city hospitals in the United States had achieved the facilities,
organization, and staff they needed to apply the rapidly advancing
scientific knowledge; others were striving toward that end. As the
poorer medical schools closed or merged with others following the
publication of the Flexner report, the remaining ones became
increasingly involved in patient care, clinical teaching, and clinical
research. City hospitals were supported and stimulated by these
activities of the schools, and usually the stronger the ties between
school and hospital the better both institutions fared. For ideal
cooperation between medical schools and hospitals, secure and stable
bonds were essential—bonds that assured unified action in caring for
patients while allowing each partner a maximum of freedom in other
functions. Such bonds were established in city hospitals by means of
university-type services. Such a service is directed by one chief, a
senior faculty member in an affiliated medical school. He and the
attending physicians or surgeons and house officers are appointed by
agreement between medical school and hospital, and the doctors and
medical students are responsible to the chief in all professional
medical matters. Supervision of medical care, teaching, and research is
controlled by this staff, subject only to the general rules of the
hospital.

Development of university services often required skillful negotiations and sustained effort. Harvard Medical School, for instance, had
to tread a long and tortuous path at the Boston City Hospital before

finally achieving a university service. Harvard faculty members had been prominent on the staff of the city hospital, and its students had been taught there from the start. But throughout the nineteenth century the medical staff was a loose confederacy of doctors, each responsible for the medical care of a few patients for part of a year. The techniques and even the basic principles of treatment could vary depending on the doctor in charge of a ward at a particular time. General rules of the staff related to conduct rather than medical care. Staff opinion also decided who would be recommended for promotion, and these recommendations were usually based on seniority.[1]

Meanwhile medical science was making it possible to treat patients in the hospital more safely (through knowledge of how infections are transmitted and the introduction of antisepsis, then asepsis), more comfortably (by the use of anesthesia and the newer drugs for sleep, sedation, and the relief of pain), more exactly (through X-ray examinations, counts of blood cells, chemical tests of blood and excreta), and more effectively (by the use of recently introduced specifics, such as diphtheria antitoxin, insulin, and thyroid, and by newer surgical techniques and innovative operations).[2] These advances speeded the changeover in city hospitals from a caretaking to a dynamic philosophy. Patients came flocking to these hospitals as well as to others. All these changes demanded a more effective organization of the medical staff.

University leaders, seeking to improve medical education, saw the need for better organization of teaching in hospitals. President Charles W. Eliot was astonished to find that in 1888 there was "not a single hospital infirmary or dispensary" in which Harvard had the least control over appointments. He and forward-looking members of the Harvard faculty realized that teaching could not be successfully coordinated with medical care and worthwhile research could not be performed unless policies and planning within a service were kept consistent by the appointment of a long-term director. But it was not until 1908, when word came that Harvard contemplated establishing a teaching hospital of its own, that the staff of the Massachusetts General Hospital was willing to establish a continuous service under a single chief in each of the two major services, medicine and surgery. Three years later the staff accepted Harvard's proposal of an outsider as chief of a medical service. He was David L. Edsall, educated in Philadelphia and at the time of his appointment professor of medicine at Washington University in St. Louis. When the Peter Bent Brigham Hospital opened in 1912, the divisions of medicine and surgery were placed in the charge of Harvard professors from the start.[3]

Bellevue Hospital had recently reorganized its staff so that each of three medical schools was to appoint a chief of a medical and a chief of a surgical service to be director for the entire year and to nominate staff doctors for his service. Following its example, Boston City Hospital worked out a compromise in 1915 whereby one of the medical services was allocated to Tufts and another to Harvard Medical School, each to be directed by a chief nominated by the schools.[4] Even then the schools' hands were tied; they were not free to nominate an outsider. In 1899 Eliot had tried to bring William S. Thayer from Johns Hopkins to be Harvard's senior professor at the city hospital. When the majority of the attending staff lined up against the appointment because it would have violated their seniority system, Eliot was forced to accept an insider, George G. Sears, instead.[5]

The real beginning of university services at Boston City Hospital can be dated from 1923, when the Thorndike Memorial Laboratory opened. Some years before George L. Thorndike had bequeathed a fund to the hospital in memory of his brother, a former attending surgeon at the hospital; this bequest was used to erect a research laboratory in his name. Harvard had available just the man to direct it: Francis W. Peabody, a promising young physician who had spent his postgraduate years acquiring skills in the new discipline of clinical investigation at Johns Hopkins, the Rockefeller Institute for Medical Research, and in Europe. Probably because Peabody was a Bostonian and a graduate of Harvard Medical School and had been a house officer at the Massachusetts General and Peter Bent Brigham Hospitals, the city hospital staff did not object when in 1921 he was appointed director of the Thorndike Laboratory and chief of the Harvard medical services.[6] The research laboratory quickly became the hub of the Harvard medical unit; ideas and action began to flow freely from center to periphery and periphery to center until, as in a wheel, it was impossible to say which was the support and which the supported. (For a further discussion of the Thorndike Memorial Laboratory, see below.)

For a unit such as this, where teaching, research, and patient care were intertwined, the funds came from several sources and were contributed in different proportions at different times. In 1941 the Harvard Medical Unit at the Boston City Hospital had a staff of twelve full-time doctors in addition to fellows, resident physicians, interns, and technicians. Some of the staff doctors were paid by the hospital, some by the university, and some were paid jointly by the hospital and medical school, including four of the senior staff. Conversely, all professional salaries in the neurological unit, which was also a joint enterprise of Harvard and the hospital, were paid by Harvard University.[7]

Establishment of university services in surgery at the Boston City

Hospital generated more conflict and took much longer than establishing the medical unit. Harvard doctors had taught surgery to medical students from the opening of the hospital and in 1893 were joined by faculty and students from Tufts Medical School. But the old guard on the hospital's surgical staff, which included many doctors unaffiiated with medical schools, doggedly opposed control of any surgical services by medical schools. In the 1890s they blocked Harvard's attempts to establish a continuous surgical service under a single chief. Finally, in 1928, after repeated efforts, Harvard was granted a partial substitute, a teaching service. The chief of the service was to be elected by the surgical staff in the usual manner, on the basis of seniority, and an associate chief, nominated by the university and approved by the staff, could be placed in charge of teaching and research but was strictly confined to those areas.[8] On such a service it is impossible to ensure optimal medical care, teaching, and research. A chief who has risen by seniority is often limited in vision and interests, while the authority of the associate chief is sharply restricted, interfering with administration and impeding recruitment of promising younger doctors. Nevertheless, because this plan represented a partial step toward a full university service, Harvard was willing to try. The Fifth Surgical Service was designated a Harvard teaching service, and Irving J. Walker, a highly respected clinical professor of surgery at Harvard, who had risen through seniority at the city hospital and who divided his time between that hospital and private practice, was appointed chief. The brilliant Edward D. Churchill, associate professor of surgery at Harvard, was made full-time associate chief of the service and director of the hospital's surgical research laboratory. In the same year another surgical service was designated a Tufts teaching service, and one of the school's professors, Horace Binney, was appointed, on a seniority basis, as year-round chief.[9]

Thus, the hospital sanctioned university sponsorship of teaching on surgical services while continuing to reject the universities as full partners in the direction of the services themselves. When Harvard requested permission to nominate an occasional outsider who was especially proficient in teaching, the surgical staff replied by recording itself as opposed, in general, to staff nominations by outside institutions.[10] In 1930, after the hospital staff had again blocked a plan to create true university services for each of the three Boston medical schools, Churchill returned to the Massachusetts General Hospital, eventually becoming its distinguished chief of surgery and full professor at Harvard. After another attempt to establish a full-time university service failed in 1942, Harvard abandoned the teaching of surgery at the city hospital.[11]

In 1949 the American Medical Association withdrew approval of the fourth year of the surgical residency at the hospital, thus making it impossible to offer a complete training in surgery.[12] Alarmed, the surgical staff recommended that Boston University, Harvard, and Tufts each be invited to accept a surgical service, but what they offered with one hand they withdrew with the other: they still insisted that the chief of a service be selected according to seniority and that surgeons assigned by the schools be subject to the seniority system. For instance, Eugene E. O'Neill, who was made chief of the Tufts surgical service in 1951, found his appointments blocked by the hospital's chief surgeons. Finally, after many bitter meetings, the surgical staff yielded in order to preserve their residency training programs. They agreed in 1955 that the director of one surgical service would be nominated by each affiliated medical school (Boston, Tufts, and Harvard) and that he would nominate his staff directly to the trustees. Thus, Harvard finally established a full-time university surgical unit, and in 1955 J. Engelbert Dunphy was appointed director of the Fifth Surgical Service and of the new Sears Laboratory for Surgical Research. After half a century Harvard Medical School had a surgical unit at the city hospital comparable to its services at its other main teaching hospitals.[13]

The arrangements negotiated by Dean George P. Berry are worth giving in detail, because they demonstrate how support for a university service at a city hospital must be gathered from various sources. In 1930 or 1931 Charles H. Tyler had bequeathed over a million dollars to the hospital to support a laboratory for surgical research in memory of his friend and physician George G. Sears, a former Harvard professor of medicine and an attending physician and member of the board of trustees of the Boston City Hospital. After many discussions among the trustees, Mayor John Hynes, and Dean Berry, it was agreed that Harvard would give $15,000 annually from its endowment income and the hospital trustees would designate $15,000 annually, in addition to the $20,000 already provided toward Dunphy's salary, and that $30,000 would be supplied from the income of the Tyler Fund. Thus, at the same time a director of the endowed surgical laboratory and a chief of the Harvard surgical services were assured, and funds were provided to run the laboratory and the service.[14] Tufts and Boston University Medical Schools also developed surgical units with chiefs and assistants selected by them, who were full time or essentially full time, and associated research laboratories.

In contrast, university services were established more smoothly in Baltimore's city hospital, probably because of the example of Johns Hopkins Hospital a few miles away. At the turn of the century Bay

View Asylum had consisted mainly of an almshouse, a tuberculosis sanatorium, and an insane asylum. Yet there were enough patients with the general run of medical and surgical illnesses to interest the medical schools. In 1891 the University of Maryland, the College of Physicians and Surgeons, and Johns Hopkins University had pledged themselves to supply attending physicians each day and a surgeon for every important operation. The first two schools sent their students for teaching also, while Johns Hopkins embraced Bay View only where its own resources were scanty. As early as 1886 it began supplying attending and resident physicians to the psychiatric wards and a few years later sent medical students there, and in 1898 a Hopkins associate professor was placed in charge of these wards. Likewise, William H. Welch, whose national reputation had brought about his appointment to the chair of pathology at Hopkins, found that a drop in the university's income had delayed the opening of its hospital and left him "somewhat stranded as regards medical teaching and human autopsies." Thus, when his protégé, William T. Councilman, was added to the staff at Bay View, the abundant autopsy specimens he brought to Hopkins were most welcome.[15]

Between 1884 and 1892 Johns Hopkins University appointed three outstanding department heads who became pathologist-in-chief, physician-in-chief, and surgeon-in-chief, respectively, of the Johns Hopkins Hospital: Welch, Osler, and Halsted. (Halsted was appointed acting chief at first because the university was not sure that he had fully recovered from his addiction to morphine.)[16] Although these men were never members of the staff of the city hospital, they had a profound, if indirect, influence upon it. In 1910, when the new general hospital at Bay View was about to open, the Board of Supervisors of City Charities asked Hopkins to nominate chiefs for these three departments. Apparently, nominations had been requested from Maryland also, and the two faculties must have consulted each other, because the Hopkins faculty nominated Thomas R. Boggs and Milton C. Winternitz of their school as chiefs of medicine and pathology and Arthur M. Shipley of Maryland as chief of surgery; these three were then appointed.[17] Obviously, Johns Hopkins had already adapted itself to the customary method of getting things done in Baltimore: a few influential persons made a decision for the good of the city and quietly implemented it through a web of social, business, and political relationships.

These three appointments were significant for the future of the city hospital. The high quality of the three chiefs, the wide-ranging reputations of Boggs and Shipley, and the association with the world-renowned departments of medicine and pathology at Johns Hopkins gave luster to Bay View. In fact, their prestige was such that hospital authorities and the affiliated medical schools were willing to overlook the absence of other essential departments, such as obstetrics, pediatrics, and roentgenology (see Chapter 5) as well as the scarcity of equipment usually deemed essential for teaching hospitals. These appointments established the custom that the chiefs of medicine and pathology at the city hospital would be members of the Hopkins faculty and the chief of surgery would be drawn from the Maryland faculty, along with the chief of the tuberculosis service, whom that school had been nominating for some time. This unwritten agreement was honored for the next forty-five years. The universities nominated the attending staff as well as the chiefs, and the chiefs nominated the resident staff and directed teaching and research as well as medical care. Thus, in contrast to Boston, Baltimore had achieved university services in one step and without strife.

Since the future of medical care at the city hospital was to depend so heavily upon the three new department heads, it is fortunate that the schools chose so wisely. Boggs had been an early graduate of Johns Hopkins Medical School and had risen to the post of chief of the clinical laboratory under Osler. In 1908 he became associate in medicine and resident physician at Johns Hopkins Hospital and was made associate professor when he began his career at Bay View in 1911. When Osler had left Hopkins in 1905, he had handed Boggs his stethoscope with the injunction, "Now carry on my work." Boggs made the execution of this command his life's goal. (Boggs carried his idolization of Osler to an extreme. After he had returned from a trip to England, I was present when he told his house staff, "I will not visit Oxford, because Dr.———there missed Osler's empyema.") As a perceptive, skillful, and wise consultant, as a knowledgeable, scholarly, yet kindly and charming bedside teacher, he followed the style and methods of the master as closely as any of Osler's disciples. His daily clinics at the bedside attracted house officers in abundance, and his wide acquaintance among leaders in internal medicine accounted for much of the city hospital's national reputation.[18]

Shipley was a product of the University of Maryland both as student and resident in surgery. He rapidly became a skillful diagnostician and operator. These talents, together with his leadership ability and genius for organization, accounted for his rapid rise to the professorship of surgery at his alma mater and his appointment as surgeon-in-chief at Bay View. He also managed to find time for a busy private practice. This blend of practice with teaching and the exploration of new procedures brought superior young men to his training programs at the University of Maryland and at the city hospital.[19]

The third newly appointed chief, Milton C. Winternitz, a dynamic and forceful teacher, was soon tempted away from Bay View by Yale

University, where he became professor of pathology and later dean of the medical school. But just as Winternitz followed several excellent pathologists at Bay View (see Chapter 3), so was he succeeded by other able ones, including James R. Cash and Wiley D. Forbus, who later became professors of pathology at the University of Virginia and at Duke, respectively. Until William G. MacCallum retired as chief of the department of pathology at Johns Hopkins in 1941, the departments at Hopkins and the city hospital worked together to the advantage of both.[20]

Yet, the hospital's ties with Johns Hopkins Medical School were few. (For instance, between 1911 and 1921 only one reference to the city hospital appeared in the minutes of the Johns Hopkins medical faculty.) The stimulus for closer relationships came from George Walker, who had been so active in acquainting the ruling elite of Baltimore with the importance of the hospital. In 1932, when the hospital's superintendent suggested that the University of Maryland medical school move to the city hospital grounds and that the hospital affiliate with Maryland alone, Walker warned Dean Alan Chesney of Johns Hopkins that this could happen if his school did not resolve to participate more thoroughly in the hospital's affairs. Chesney roused department heads, and the professors of medicine and pediatrics began assigning more faculty members and students to the city hospitals; later, other departments became more involved. Ties with medical schools were further strengthened in 1934 with the organization of the Medical Advisory Board, which determined professional policy for the hospital within the limitations of the budget.[21]

The original chiefs had succeeded in directing their services by spending only part of their time at the city hospital. This was the custom around the turn of the century, when leading clinicians were said to give "their mornings to the poor and their afternoons to the rich." As hospital services became larger, administration more complex, and travel time longer, the combination became more difficult. The medical schools pressed the hospital to obtain funds for full-time chiefs of clinical services, first in pediatrics in 1945 and over the next two decades in other disciplines. Boston had preceded Baltimore by twenty to forty years in installing full-time chiefs, and these chiefs had built staffs of superior clinicians and investigators, who in turn attracted superior house officers and fellows and gained a worldwide reputation for the hospital. Once Baltimore's city hospital took the same step, it began building services comparable to those in university hospitals and, like its Boston counterpart, attracted outstanding doctors.[22] The caliber of the doctors recruited to the full-time staff can be judged by the positions they held later: George S. Mirick became professor of medicine at New York University and later the Medical

College of New Jersey; Francis Chinard, professor of medicine at the Medical College of New Jersey; Julius Krevans, dean of medical affairs at Johns Hopkins Medical School and later dean at the University of California at San Francisco; Charles C.J. Carpenter, professor of medicine at Case-Western Reserve University; and Peter Safar, professor of anesthesiology at the University of Pittsburgh.[23]

Some city hospitals reached satisfactory agreements with universities much later. Cooperation was not easily achieved in every case, even though other hospitals had shown the way. The road to a proper agreement was especially long at the city hospital in the nation's capital. As late as 1914 one part-time attending physician, one full-time resident physician, and three medical students who worked at night tried to care for over two hundred patients. In that year sixty consultants were appointed in the various specialties to treat patients with the more complex illnesses. Most of them made their assigned visits regularly during the school year, less often during the summer. Once, they even joined together to pay the salary of a pathologist until an appropriation could by obtained.[24]

In Washington neither Georgetown nor George Washington University had shown much interest in a hospital composed of several small buildings on a remote hillside in the southeastern section of the city, but when an up-to-date medical-surgical building accommodating 270 patients was completed in 1929, both were eager to staff the services and teach students there. The two medical schools provided an attending staff of 139 doctors, and for the first time the hospital was accredited without qualification by the American College of Surgeons.[25] Chiefs of medical and psychiatric divisions were supplied by Georgetown University, chiefs of surgery and obstetrics-gynecology by George Washington. Under each served an equal number of attending and associate attending physicians nominated by each medical school. Alternate patients were assigned to a Georgetown or a George Washington service.[26]

These were university services on paper only. The chiefs were mostly absentee supervisors; each was also chief of his department in his medical school and university hospital and had besides a busy private practice. Nor could a chief from one medical school maintain adequate control over doctors from another. Uninterested staff doctors could skip rounds or shorten them so as to render them meaningless, and even the conscientious ones had little control over the actions of house officers. Finally, medical students were not integrated into the services but were only suffered to enter the wards for three hours each morning to examine patients.

When I joined the attending staff of Gallinger in 1934, assigned by George Washington University, I received a letter from the assistant superintendent of the hospital stating that the chief of the service, a professor of medicine at Georgetown University whom I had never met, had assigned me to visit a certain ward on Tuesday, Thursday, and Saturday of each week for three months. When I arrived at the hospital on the first Tuesday, I had to hunt up the intern on that ward so that I could be informed about the patients who would be under my care. Sometimes an intern would be busy elsewhere when the attending physician arrived, in which case there was little the attending doctor could do except fume and hope the intern returned soon or else go through the motions of making rounds on his own. If an intern believed he was not learning from a particular attending physician he might find urgent business elsewhere in the hospital whenever that doctor was due to arrive. In defense of such actions, if the attending physician was poorly informed or a poor teacher, if his methods were out of date, his attendance irregular, or his rounds too hasty, the house officer had no one at hand to whom he could complain. Having just come from the Johns Hopkins and Harvard medical services at Baltimore and Boston City Hospitals, where custom required obedience and discipline, I was horrified by this desultory nonsystem.

The city hospital in Boston had established university services by installing full-time chiefs and associates and a recognized hierarchy of house officers and clinical clerks. In Baltimore and Washington the medical schools did not have to contend with a previously established staff maintained by seniority. The Baltimore hospital introduced university services by paying salaries to chiefs of services for devoting specific hours to the city hospital and by maintaining a disciplined corps of house officers responsible to the chief. Medical schools in Washington had gained the privilege of teaching their students in the city hospital in return for supplying attending physicians of their own choosing, but in 1937, when the city hospital was placed under the Health Department, a dispute arose as to who was actually in charge of medical care.

George C. Ruhland, who had distinguished himself by developing modern, aggressive public health programs in Milwaukee and Syracuse, had been appointed director of the Health Department of the District of Columbia in 1934. The vigor and widened scope of this department soon became apparent; thus, it seemed logical that Ruhland could transform Gallinger Hospital also.[27] The new health officer, comprehending the deficiencies in the supervision of medical care at the hospital, searched for and appointed five full-time chief medical officers to head the medical, surgical, obstetrical-gynecological, pediatrics, and tuberculosis services.[28] These doctors had been trained at leading medical centers and were comparable to assistant professors on a medical school faculty. The health officer had cogent reasons for his actions. The attending physicians were busy practitioners: they sometimes scanted their duties at the city hospital; they were not always willing to make a long trip across town at night to see an acutely ill patient. The system could work optimally only when an attending physician was a competent doctor and teacher and spent considerable time on patient care and in teaching house officers.

However, the health officer's new system collided with the programs of the medical schools. Having only small hospitals of their own, they were dependent upon their services at Gallinger and had invested manpower in them with the understanding that they would select the staff. The chiefs and attending doctors they nominated had been generally competent and experienced.[29] Importation of medical officers chosen by the health officer to be in charge of the services violated the unwritten agreement between schools and hospital and could be contrary to the interests of the patients when a department head in a medical school was supplanted by a doctor many years his junior. In 1939, 130 doctors from the two schools had made forty-six hundred attending rounds at Gallinger. Without their services, the cheap help of the medical students, and the prestige of the schools to attract house officers, the hospital might have reverted to the poorhouse hospital of the nineteenth century despite its handsome new building. The large corps of part-time doctors was needed, but also required was better on-the-spot supervision.[30]

Fortunately, the two groups achieved a workable compromise: it was arranged that the new chief medical officers would have supervisory duties under the direction of the chiefs of services appointed by the schools and the schools would pay the salary of a full-time chief of staff, who would coordinate the activities of attending staff, house officers, and medical students.[31] Once this agreement was reached, university deans and presidents, city officials, and meddling congressmen, assuming that all was solved, withdrew, leaving the troops in the field to work out a peace agreement. The troops succeeded, but it took time.

Before the advent of the chief medical officers, Hugh Hussey and I had begun to build the Georgetown and George Washington medical divisions, respectively, so that they consisted of a nucleus of salaried physicians, who served continuously, and a larger group of volunteer physicians, who attended for a few months each year. By the late 1940s these divisions had become effective units for medical care, teaching, and research. Each division recruited its own resident physicians and interns and incorporated medical students fully into the team as clinical clerks. Surgical services, organized on a similar model, followed some years later. These de facto university services gradually and imperceptibly came to be actual university services, recognized by the administration of the hospital as responsible for the care of patients and the training of house officers on their wards. Each service developed its own modus operandi with the corresponding chief medical officer, and as the hospital employed additional salaried doctors, they were incorporated into one of the university services.

Departments such as obstetrics-gynecology and pediatrics were run by the chief medical officers, who were more or less formally deputized by the university-nominated chiefs of these services. These services also worked effectively, partly because the first chief medical officers set a pattern of devotion and ability, and partly because the members of the corresponding departments in the universities were too busy elsewhere to be concerned with the reins of power at the city hospital. Thus, a working relationship between medical schools and hospital, which included some university services, had evolved by the late forties, although the schools did not obtain a written agreement with the Health Department until 1958.[32]

The city hospitals of Boston, Baltimore, and Washington achieved university services eventually, in different ways, and at different speeds. But in Chicago's city hospital the authorities steadfastly resisted valid and effective affiliations with medical schools. At the end of the nineteenth century some had considered Cook County Hospital one of the best hospitals in the Midwest and one of the leading city hospitals in the country. An internship there was highly prized, and nearly every attending physician or surgeon was on the faculty of one of the medical schools. Yet, well into the twentieth century an inefficient staff organization remained in which each attending doctor was an independent peer of the realm, in complete charge of "his" ward and accountable to no one for his actions.

Cook County Hospital first tried to meet the need for minimal standards, supervision, and a modicum of uniformity by establishing a hospitalwide efficiency committee, which rated the performance of each attending physician and his staff by surveying the records of patients under their charge. But judgment was reported long after the event, the committee's ratings were not followed by action, and the efficiency ratings became a subject of derision.[33] A more direct system of supervision was needed.

In 1931 the Chicago Medical Society proposed that divisions be assigned to each of the four accredited medical schools in Chicago, each division chief to be appointed by the county commissioners on the school's nomination. This proposal died of malnutrition when hospital authorities ignored it, and each medical school had to be satisfied with having its faculty members "capture and retain their share of staff appointments" by passing civil service examinations, held every six years.[34]

A survey committee of the American College of Surgeons, reporting in 1932, failed to grasp the nettle of affiliation with medical schools, although it pointed out the lack of organization within the hospital staff and the inadequate supervision of house officers. The committee could find "no by-laws, rules, and regulations setting forth the organization, duties, relations, responsibilities, and policies of the various units of this large institution." Thus, a rule of men prevailed (and individual men at that), not a rule of law. The surveyors criticized the lack of associate attending physicians to help attending doctors supervise interns and commented that too much responsibility was placed on the intern when the attending doctor was not able to make thorough daily rounds.[35] Furthermore, they could find no evidence that the clinical work of the hospital was systematically reviewed and analyzed.[36]

A few years later a citizens' committee reported the same defects and again recommended allocation of services to medical schools. When publicity generated by this report forced the commissioners to give an account of their stewardship, they blandly claimed that division into university services had been accomplished "by placing the successful candidates of the same schools in grouped wards so that they may advantageously work together."[37] This, of course, gave the schools no authority and thus no effective means of influencing policies, raising standards, improving patient care, or training house officers.

The politicians who ran the hospital had one guiding principle: get all the power you can and never give any away. University-controlled services meant to them a loss of power. Their tactics of opposition to change included fulsome praise of the existing hospital, its former great men, and its existing staff. Since the system of civil service examinations for staff positions was working in their favor, they extolled its advantages by contrasting it to the ancient disputes among the schools that had led to their ouster in 1882 (see Chapter 3). Actually, by the 1950s the civil service examinations were not highly selective, since sixty of one hundred points were based on experience, on the hospital staff and elsewhere, and because so few doctors now took the examinations that the hospital was reduced to appointing everyone who received a minimally acceptable grade.[38]

While hospital authorities boasted of the caliber of the house staff, fewer candidates were appearing. For many years the hospital had accepted only graduates of fully accredited medical schools in Chicago. In 1941 it welcomed graduates from similar schools elsewhere, and finally it was forced to use aggressive recruiting measures, such as sending teams of doctors to medical schools throughout the country. Yet, the superlatives of the hospital's recruiters and the splendor of its showcase units contrasted so sharply with the dingy, dirty, crowded wards and the overworked staff that the need for good house officers far exceeded the supply. By 1967 approximately 7 percent of the hospital's internships and residencies were vacant, and 35 percent were filled with graduates of foreign medical schools. In pediatrics 88 percent of house officers were foreign graduates.[39]

Conscientious house officers, finding they could not get enough help from attending doctors on their own services, would ask clinicians from other services to give advice on their sicker patients. For several decades the most sought-after medical consultant was Edmund M. Foley. He began his long career on the attending staff of Cook County Hospital and the University of Illinois faculty in 1922 and retired in 1964.[40] His broad knowledge of medicine, his penetrating analyses of the most difficult diagnostic problems, and his method of driving home a lesson with a bit of appropriate humor were matched by his kindness and humanity toward patient, student, and doctor alike. As president of the staff and chief of the University of Illinois medical service for many years, he saw his mission as enabling administrators and staff to work together. Only through the dedication of Foley and others interested in the welfare of the patients could a patched-together hospital carry its enormous load without breaking down sooner than it did.

Beginning in the late 1940s the departments of medicine of some associated medical schools had placed full-time faculty members at Cook County Hospital to supervise the work of medical students. Since teaching students required that medical care measure up to accepted standards, the schools were forced to form a staff to coordinate the activities of attending physicians, house officers, and students. Even though the chiefs of the medical school divisions had no authority over attending physicians, they accomplished their mission through the delegated authority of the department head in the associated medical school and by agreement among department heads of the four medical schools involved: Chicago Medical School, University of Illinois, and Loyola and Northwestern Universities. (As late as 1967 there was still no formal agreement between any medical school and Cook County Hospital; eventually, in 1975, the hospital signed an agreement with the University of Illinois, the principal teaching school at that time.) Thus, teaching services were established in medicine and surgery with the tacit consent of hospital authorities but without a written agreement, without specific representation on hospital councils, and thus without real power. The entire system could be abrogated on a moment's notice. Cook County Hospital in the 1950s had reached the status achieved at Bellevue over half a century before, which Flexner in his 1910 report had censured as inadequate for proper medical education.[41]

Hospital authorities tacitly admitted the need for supervision by building up their own corps of full-time doctors to supervise and teach the house staff. Thus, they flattered the medical schools by imitating their system while they complicated the schools' and their own tasks by setting up independent groups with many of the same objectives. By 1963 the full-time attending staff numbered forty-five.[42] In the past the medical staff of Cook County Hospital had been a multitude of individual doctors striving to build empires; in the 1950s it was a multitude of groups striving to build empires.

During the half century following 1910 medical schools in Boston, Baltimore, Washington, and other cities built satisfactory university services in city hospitals by combining volunteer attending physicians with doctors whose salaries came from hospital, university, or both. Hospital authorities accepted university services so long as the cost to them was minimal, because this system took administrative burdens off their shoulders, because it helped appease their critics, and because they recognized that medical students and superior doctors improved medical care. In contrast, those who governed Cook County Hospital, fearing a loss of power, blocked effective university services and thus accelerated the hospital's downward slide.

The histories of these four hospitals illustrate obstacles to the evolution of university services in city hospitals and how they could be overcome. Throughout the nation some medical schools were closely, others loosely, affiliated with city hospitals. But in 1940, of sixty-seven medical schools reporting to the American Medical Association, only eighteen were not associated at all with a city hospital, and twelve of these were public schools that utilized as their main teaching facility a state-owned general hospital located in a principal city (see Table 8). The remaining public school, at federally owned Howard University, used federally owned Freedmen's Hospital for teaching. Thus, among the sixty-seven medical schools, only five, all private, were associated with neither a city hospital nor a state- or federally owned hospital that functioned as a city hospital.

Of the fourteen public medical schools associated with city or county hospitals, nine were the only medical schools at these hospitals, six utilizing the city hospital as their main teaching hospital. A smaller proportion of private medical schools were the only schools affiliated with their respective city hospitals, while fourteen of the private schools had to be content to share a hospital with two or more schools. Among eighty-three medical schools reporting in 1965, the proportion of schools not associated with city hospitals and those associated with two or more other schools had changed little. The number of those associated in a city hospital with one other medical school had fallen from 24 to 10 percent of all schools, and those that

**Table 8. Association of medical schools with city and county hospitals.**

| Nature of association with a city or county hospital | Ownership of medical school | | | | | |
| --- | --- | --- | --- | --- | --- | --- |
| | Public[a] | | Private | | All schools | |
| | 1940 | 1965 | 1940 | 1965 | 1940 | 1965 |
| Not associated | 13 | 16 | 5 | 5 | 18 | 21 |
| Associated as the only medical school | 9 | 22 | 9 | 11 | 18 | 33 |
| Associated along with one other medical school | 4 | 2 | 12 | 6 | 16 | 8 |
| Associated along with two or more other medical schools | 1 | 4 | 14 | 17 | 15 | 21 |
| Total | 27 | 44 | 40 | 39 | 67 | 83 |

*Sources:* American Medical Association, Council on Medical Education and Hospitals, "Medical Education in the United States and Canada," *Journal of the American Medical Association* 117 (August 30, 1941):686–687; American Medical Association, Council on Medical Education and Hospitals, "Hospital Service in the United States," ibid. 116 (March 15, 1941):1083–1142; American Medical Association, Council on Medical Education, "Medical Education in the United States," ibid. 198 (November 21, 1966):852–853; *Hospitals,* Guide Issue 40 (August 1, 1966):18–244.

a. Owned by a city, state, or the federal government.

were the sole school utilizing a city hospital had increased from 27 to 40 percent of the total.

The coexistence of two or more university services in a hospital was bound to generate rivalries. Some of these were healthy. Pride in one's service goaded staff and students to give the extra ounce of effort that made the difference between acceptable and superior medical care. A worthwhile innovation on one service was often copied by another. Doctors trained on one service, finding no empty niche above them, could move to a desirable position on another.

An example of a highly successful service in a hospital associated with several schools was the Harvard medical unit at the Boston City Hospital, which was directed from 1921 on by men of international reputation, such as Francis W. Peabody, George R. Minot (a Nobel laureate), William B. Castle, and Maxwell Finland, and which trained 922 resident physicians and fellows, up to 47 percent of whom later achieved professorial rank in medical schools in the United States and throughout the world. Few if any modern medical institutions can equal this record. Throughout this unit's existence its members were imbued with the philosophy that patient care, teaching, and research should exist side by side and that the spirit of one should permeate the other, that doctors who manage patients, who teach students, or who inquire into the whys and hows of nature—should all be curious, innovative, and probing; should all be trustworthy and conscientious; should all be compassionate and caring.[43]

Tufts and Boston University also developed services of high caliber at the city hospital. Opinions differed, however, regarding the value of their association with the Harvard services in the same hospital. Experts associated with one school were a valuable source of information, inspiration, and advice to the staffs of other schools, but the presence of a strong unit in a special area controlled by one school could inhibit the development of much-needed expertise in the other schools. If, on the other hand, all three schools developed units in a particular area (as occurred at the Boston City Hospital as the Tufts and Boston University units became larger), the duplication increased costs and complicated administration.[44] This was one reason why authorities at Boston City Hospital eventually decided that the hospital should affiliate with only one school (see Chapter 8).

Competition between medical schools in a city hospital could also be destructive, consuming time, delaying important programs, and generating bitterness. An example was the controversy over control of the surgical service at the Baltimore City Hospitals. From 1911 on each medical school nominated chiefs of certain clinical services. By 1938 these were surgery, obstetrics, and tuberculosis by the University of Maryland and medicine, pediatrics, and psychiatry by Johns Hopkins. The chief chose attending physicians largely or entirely from the faculty of his school and arranged for them to teach its medical students. In the early 1950s arrangements were made for Maryland to nominate attending physicians on the medical service to teach their students and for Hopkins to nominate some of the surgical staff.[45]

After Shipley retired as chief of surgery in 1939, his successor at the University of Maryland concentrated on building his department at the university hospital and assigned others to run the city hospital service. Some members of the staff believed the service had become less effective under this program. By the 1950s, when there were full-time chiefs of eight other major departments at the hospital, the chief of surgery retained his private practice. The surgical service decreased greatly in size, the number of acutely ill patients fell from a high of 130 to thirty-five, and the Hopkins portion of the surgical staff felt strongly that leadership of the surgical service was not vigorous enough.[46]

When the surgeon-in-chief's post became vacant in 1955, a well-trained and dynamic surgeon, Mark M. Ravitch, was among the Hopkins surgeons at the city hospital. Members of his unit insisted that he be appointed chief instead of Maryland's nominee. When the schools could not agree after many attempts to resolve the controversy and a public airing in the newspapers, the Department of Public Welfare appointed Ravitch. As anticipated, he built up a superior surgical service, but so much bitterness had developed between the two faculties that the University of Maryland eventually withdrew entirely from the city hospital.[47]

The two universities had worked together harmoniously for half a century by means of a tacit understanding, but this had proved too fragile. The remark of the city's health commissioner that "a primadonna...got over into the other fellow's pasture and stole his cow" belied the significance of the episode. The withdrawal of the University of Maryland, while it lessened friction between the medical schools, tended to isolate further from the community a city hospital that was already little known by the people of Baltimore. The hospital was now hitched to the Hopkins star and it depended upon that school more than ever, while the school, in turn, was committed to considering the welfare of the city hospital in all its actions. The nagging question remained: Did the short-term advantage to the hospital of acquiring a superior department head outweigh the long-term value of support from two medical schools?[48]

Another example of a potentially destructive rivalry was the competition, initiated by Huey Long, between Tulane and Louisiana State Universities for teaching facilities at Charity Hospital (see Chapter 5). Fortunately, relations between the two medical schools improved after Long's death and after expansion of the hospital provided adequate facilities for both schools.[49]

Of course, the most practical arrangement was for one medical school to be affiliated with one city hospital. Such an affiliation was especially satisfactory when the city hospital became in effect the university hospital. As early as 1916 the city of Cincinnati placed control of the medical and nursing divisions of the city hospital under the University of Cincinnati, and this close connection has continued. The University of Rochester developed an arrangement between its medical school and the city hospital that was said to be advantageous to both: the university and city hospitals were united physically; the medical school controlled both and contracted with the city for the care of city hospital patients.[50]

Yet, affiliation with medical schools was not a panacea for everything that ailed city hospitals. This was demonstrated in 1961 in New York City, when a crash program was begun to affiliate all municipal hospitals with a medical school or a voluntary hospital. By 1964 the city government had shifted control of ten municipal general-care hospitals into the hands of nongovernmental institutions, two of which were medical schools and one a hospital in the process of developing a medical school (Mount Sinai). The other affiliating institutions were voluntary hospitals, most but not all affiliated with a medical school.[51] These affiliations were entered into because New York's municipal hospitals in 1960 were typical of those in other large cities: many were obsolescent, even unsafe; all were poorly maintained; there were continuous shortages of nurses, other personnel, and essential equipment; diagnostic and outpatient facilities were inadequate; and the Department of Hospitals was "hamstrung by rigid and overcentralized budget, purchasing and personnel procedures." Dr. Ray E. Trussell, commissioner of hospitals from March 1961 to July 1965, pushed the idea of affiliation and negotiated the agreements. Responsibility was transferred to the affiliating hospitals or medical schools for "directing, reorganizing and staffing programs of patient care and house staff training." The agreements emphasized "full-time directors of all major services supported by well-qualified, full-time and part-time staffing. . . operating under the supervision" of the affiliating institutions. Funds for these services, provided by the city and spent under the direction of the affiliating institution, amounted to $50 million in the 1965–1966 budget.[52]

The greatest improvements made by the program were in the quality of doctors' services. Incompetent, inadequately trained, and unlicensed doctors were replaced by board-certified or board-eligible doctors. (Board-certified doctors are those who have successfully passed the examinations of the national specialty boards; board-eligible doctors are those who have completed approved training sufficient to admit them to these examinations.) The number of approved resident-training programs was greatly increased. The number of interns educated in unapproved foreign medical schools was reduced considerably in municipal hospitals directly affiliated with medical schools and somewhat decreased in those affiliated with voluntary hospitals that had their own affiliations with medical schools. Moreover, these improvements were made during a period of shortage of American-trained house officers. Full-time chiefs of major services were appointed to supervise and coordinate the work of attending and house staffs. Other improvements included installation of sorely needed equipment for diagnosis and treatment, construction of some new buildings, and renovation of many others. The number of patients cared for in outpatient clinics of municipal hospitals increased from less than 2.7 million to over 3.2 million between 1960 and 1965, while the number attending the clinics in voluntary hospitals decreased slightly.[53]

These gains were partially offset by several failures. Preventive measures and comprehensive care were not promoted in parallel with high-skill, specialized medical procedures. Home and outpatient care and care of the chronically ill received insufficient attention despite increasing demands. Community ties were weakened as local doctors on each hospital's voluntary staff were displaced by full- and part-time salaried doctors. Medical schools, being interested in only certain portions of the house of medicine, had done little to shore up the other parts.[54]

The lesson of New York City's experience was that affiliations with medical schools could help city hospitals by strengthening the medical staff and improving the quality of care rendered by doctors, thus indirectly benefitting some additional activities, but that these affiliations would not promote other important functions of the hospital badly in need of help. Furthermore, affiliation programs entered into hastily, as in New York, could divert interest and funds from other ailing municipal health services.

Affiliations with medical schools tended to aid city hospitals in another area: they stimulated interest in research. Before and during most of the nineteenth century research had been an individual enterprise. A typical example was William W. Gerhard's differentiation in 1836 of typhoid and typhus fevers based on his observations of patients, mostly at the Philadelphia city hospital (see Chapter 1). Research had also been individual when it concerned experiments, as when John C. Munro, a surgeon at the Boston City Hospital, dissected out the blood vessels involved in patent ductus arteriosus, a congenital abnormality of the large blood vessels. On the basis of these dissections, he described in 1907 the method by which this syndrome could be remedied by an operation.[55] When established measures failed to fit the varied and peculiar characteristics of patients in city hospitals, doctors there were accustomed to trying the new. For instance, they had been among the first to try Lister's techniques of antisepsis. George Derby used phenol (carbolic acid) on a patient with a compound fracture of the thigh at the Boston City Hospital in September 1867, soon after the publication of Lister's experiments. He applied the phenol freely with a sponge to the external wound—a technique that was probably too crude to appeal greatly to surgeons in private hospitals at that time—and the wound healed completely. According to J. Collins Warren, this was the first reported use of phenol in Boston surgery. The Boston City Hospital also pioneered in 1869 in the publication of its own journal of clinical reports. At the turn of the century it was still the only hospital in Boston maintaining such a publication.[56]

In one important field city hospitals helped break the tradition that research was an individual doctor's personal prerogative. Proper isolation of patients with contagious diseases and treatment with newly discovered methods, such as diphtheria antitoxin, required teamwork by a specially trained group of doctors and nurses. Pioneer work on antipneumococcal serums had been done at the Rockefeller Institute for Medical Research; but since large numbers of patients were needed to prove the effectiveness of the serums and test the relative value of each preparation, the scene quickly shifted to the city hospitals. Beginning in 1919 a study of these serums was begun at the Boston City Hospital, followed by others at Bellevue and Harlem Hospitals in New York City. These and subsequent studies required order, discipline, teamwork; they could not be carried out properly in the muddle and confusion of some city hospitals. They were best performed on university services of city hospitals, where both good order and large numbers of patients could be found. When the sulfonamides and antibiotics were introduced in the late 1930s and the 1940s, their major proving grounds were also the city hospitals.[57]

The potential of the city hospital for studying drugs is exemplified in the career of Maxwell Finland. After graduating from Harvard Medical School in 1926 and serving as house officer at the Boston City Hospital, he joined the Harvard Medical Unit, eventually becoming director of the unit and professor of medicine. A tireless worker, a persistent interrogator of nature, a meticulous observer and recorder, a broadly knowledgeable clinician, Finland trained 137 fellows during four decades and taught hundreds of house officers and students. His thorough studies on the actions, effectiveness, and adverse reactions of serums, sulfonamides, penicillin, and dozens of other antibiotics made him the world's foremost authority on the treatment of infectious diseases. Long hours of plowing in fertile fields brought an abundant harvest. Such is the challenge of the city hospital.[58]

In the early twentieth century some hospitals installed laboratories specifically for research. The Rockefeller Institute for Medical Research was opened in 1904 on the European model of separate research institutes; more typical in the United States was an institute established as part of a large city hospital, where specialized equipment and personnel could be used to study the many and varied diseases encountered in the hospital. Apparently the first research institute to be established as a part of a city hospital complex was the Russell Sage Institute of Pathology, built in 1907 at the city hospital on Blackwell's Island in New York. In 1912 or 1913 the institute became associated with the Cornell Medical Division at Bellevue. Here Graham Lusk, scientific director of the institute, Eugene F. DuBois, professor of medicine at Cornell, and a team of investigators pioneered in determining the metabolism in various diseases by utilizing a specially designed respiration calorimeter to measure the gases breathed in and out. Their demonstration of the harm incurred from a starvation diet in typhoid fever patients and the lifesaving value of a regimen high in protein and carbohydrates revolutionized the dietary treatment of this disease.[59]

The research institute that influenced medical care, medical education, and medical research as much as any in the United States was the Thorndike Memorial Laboratory of the Boston City Hospital. Its founding may not have been "the most important event in the latter day history of Boston," as Mayor Curley hyperbolized at its dedication in 1923, but its location assured that its investigators could not "remain remote or isolated from the problems of medical practice, for these problems must press upon them owing to the very propinquity of the general wards."[60] The Thorndike Laboratory became known especially for its contributions to clinical medicine: the development of the first practical liver extract administered by injection, elucidation of the proper treatment for cirrhosis of the liver with diets high in protein and low in sodium, visualization of the heart and great blood vessels by X-ray films following the injection of contrast dye, and studies on the absorption, metabolism, and excretion of serums, sulfonamides, and antibiotics. More basic investigations were also pursued. These included Castle's demonstration of the cause of pernicious anemia, the absence from the stomachs of patients with this disease of the factor that facilitates absorption of vitamin $B_{12}$, which builds red blood cells; and elucidation of the mechanism of synthesis and transport of thyroid hormones and the production and action of numerous other hormones by Norbert Freinkel and his associates.[61]

When the laboratory was founded, the editor of the *Boston Medical and Surgical Journal* had predicted that "the physicians and surgeons in charge of the wards cannot fail to be stimulated and benefitted by the researches that are undertaken in the laboratory." The prophecy proved correct: house officers were motivated to become superior doctors, and many became teachers and investigators. The Harvard unit also set an example for the other units in the hospital. The Tufts medical service established a research laboratory for inhalation therapy in 1942, and by 1963 Tufts had its own research building.[62] Surgical and neurological research laboratories were also established in the hospital.[63]

Cook County Hospital also had its research institute. In 1902 the John McCormick Institute for Infectious Diseases had been established nearby. After the institute closed, the county commissioners purchased its buildings and turned them over to the Hektoen Institute, which opened in 1943 in conjunction with Cook County Hospital. Although organized as a nonprofit corporation, it was apparently controlled by the hierarchy that ran Cook County Hospital. Perhaps because its policy did not include working directly with medical schools, this institute never achieved prominence in research or training investigators comparable to that of the Thorndike Laboratory, despite the erection of a handsome, ten-story glass and steel building in 1964.[64]

In New York City hospital officials responded to the age of research with imaginative planning. They recognized that information on the causes and mechanisms of chronic illnesses was scant, while patients with these diseases were usually shunted off to out-of-the-way places. They reasoned that ignorance and neglect could both be counteracted by an institute for research on chronic illness. Accordingly, they laid plans for such an institute on Welfare Island, which held several municipal hospitals for patients with long-term illnesses. In 1935 the first unit was established at Metropolitan Hospital in affiliation with Columbia University and its medical service at Bellevue Hospital. Under the direction of David Segal, studies were conducted on nephritis, cirrhosis, hypertension, and other diseases. A second unit, affiliated with New York University, was opened in 1939 under codirectors J. Murray Steele and James A. Shannon, later director of the National Institutes of Health. In the Second World War this unit tested hundreds of compounds for their effects on malaria, which was harassing Allied armies in the South Pacific. That same year both units became a part of a new hospital for chronic diseases, Welfare (later Goldwater Memorial) Hospital. These units contributed to medical education as well as research: clinical clerks were assigned by the affiliated medical schools, and many of the house officers and fellows later joined the medical school faculties or research institutes.[65]

In Baltimore the many patients in the city hospital with chronic illnesses attracted a research laboratory, but in this case its relationship to the hospital was mostly geographical. In 1940 the United States Public Health Service established a laboratory in the chronic hospital building in order to study the processes of aging. After the laboratory outgrew several expansions, in 1968 the National Institutes of Health erected a separate four-story structure on the hospital grounds. At first the laboratory studied the clinical and pathological features of aging. As these investigations opened new vistas, a comprehensive program of research evolved extending from studies of the interiors of cells to long-term observations of the physical characteristics and functions of elderly volunteers throughout the United States. This institute was not closely associated with the hospital next door, although several of its clinicians served as consultants, and a few hospital patients were studied in the course of research.[66]

In many city hospitals without resources for separate research institutions, investigations were carried out wherever space could be diverted from some other use. A notable example is the Nobel Prize-winning research by Dickinson W. Richards and André Cournand in the Columbia Medical Division at Bellevue Hospital. (Later, some of their research was moved to the Columbia University unit at Goldwater Memorial Hospital.) They developed methods for catheterizing the right ventricle of the heart and the pulmonary artery and used their techniques to study the changes caused by disease and the action of drugs on the heart.[67]

Because the illnesses encountered on their wards were so many and diverse, city hospitals were fertile fields for practical research. A valuable study of this kind was conducted at Cook County Hospital around 1932 by Joseph A. Capps and George H. Coleman. They mapped out the areas of pain sensation in the cavities surrounding the heart, lungs, and abdominal organs by inserting a wire into a cavity that contained fluid and asking the patient to indicate where he felt the sensation of pain. Their findings were immediately and directly applicable, helping doctors to make rapid and accurate diagnoses of abnormalities in the chest and abdomen.[68]

Cook County Hospital also made a useful contribution to therapy around the same time: in 1937 Bernard Fantus established the first blood bank in the United States by devising suitable methods for collecting and preserving blood and developing administrative procedures for collecting, identifying, and storing it. The success of this operation expedited the founding of blood banks in hospitals all over the world. Knowledge of infectious diseases and their therapy was advanced rapidly when the effects of serums, sulfonamides, and antibiotics were studied extensively in city hospitals, for instance, at Boston City Hospital by Finland, at Bellevue by Russell A. Cecil and others, at Harlem by Jesse G.M. Bullowa, and by me and my associates at Gallinger Municipal.[69]

The formation of university services, the establishment of a group of salaried teacher-doctors, and the conduct of research in close proximity to patients all contributed to better training of house officers. In hospitals that instituted this triad, the quality of house officers changed. They became less like hungry wolves charging from patient to patient, seeking to devour every available mouthful of information to gorge themselves for years to come, and more like scholar-practitioners, reasoning and working with members of a like community, helping patients while studying their illnesses.

The changeover from an internship lasting eighteen months to a sequence of one-year appointments, renewed until the house officer had received adequate training, accommodated both the hospital's and the house officer's needs. Under the former system the house officer's status had escalated rapidly until in the last months he had often become an instant expert, immature and cocksure, soon to be inflicted on an unsuspecting public as a practitioner and consultant. Such training suited the frenetic tempo of the city hospital but was not set at the pace of learning. It may have conformed to the sparse medical knowledge of the nineteenth century but could not encompass the widened scope of the twentieth. Under the system of one-year appointments the house officer had time to mature by carefully studying many patients and their diseases before becoming chief resident and supervising other house officers.

Of course, an ideal house officership was not achieved in every hospital, on every service, or at all times. In the hospitals I have described in detail the ideal was approached most closely on the Harvard services at the Boston City Hospital beginning soon after the opening of the Thorndike Laboratory and later upon the Tufts and Boston University services. Baltimore's city hospital professed high ideals from 1911 on, but in reality it fell short of them. Good workmen need good tools; the staff could not adequately instruct house officers without sufficient equipment. With more modern buildings and facilities and an infusion of interest and leadership from Johns Hopkins, the high objectives of the house officership were realized by the 1950s.[70]

At the city hospital in Washington, training of house officers improved progressively from 1929 and likewise reached a high level in the 1950s, when Georgetown and George Washington Universities were putting more of their resources into their services and rotating house officers from their university hospitals onto some services at the city hospital. At the same time Joseph Fazekas, as chief of staff, was shaming Congress into appropriating funds that were more nearly sufficient for the hospital's task. Cook County Hospital, true to its nineteenth-century traditions, never developed a coordinated system of well-disciplined house officerships with adequate leadership and necessary equipment and supplies. Consequently, it drew fewer and fewer of the suitable graduates of American medical schools. When the medical internship became less desirable in the 1930s, candidates continued to seek the surgical internship because of the opportunities afforded for operating on a variety of patients.[71] Such opportunities characteristically attracted surgical interns to other city hospitals also. When Jan de Hartog wrote to the newspapers, protesting against conditions in Jefferson Davis Hospital in Houston, "most of the doctors approved, only a few surgeons disagreed, as they could not see anything wrong with a hospital that provided them with so many interesting cases, and never interfered with OR [the operating room]." A surgical resident told De Hartog, "I can't see much wrong with this place. For my line of work it's a good enough training ground."[72]

Such attitudes reveal the hardening effect of continually confronting seriously ill patients, watching them endure discomforts and indignities as a result of insufficient help, deficient supplies and equipment, and a deteriorated physical plant, and seeing no hope of improving their situation. De Hartog found that some house officers reacted by acquiring a quality of "gruff and unsentimental saintliness," a few became bullies, while others turned into medical mechanics, "trying to repair pieces of human machinery allotted to them with the shabby and inadequate tools at their disposal as best they can, in tight-lipped disregard for the individual humanity of the patients." This attitude can be contrasted with the spirit prevalent on the Harvard medical services at the Boston City Hospital, expressed by Francis W. Peabody in a widely quoted speech. He warned that hospitals were likely to deteriorate into dehumanized machines and counseled students and doctors "that the secret of the care of the patient is caring for the patient."[73]

City hospitals tended to overwork their house officers—sixteen hours' work per day for weeks on end was not unusual. Despite this and the handicaps to good medical care, for a long time city hospitals were successful in recruitment because the number of medical school graduates exceeded the number of desirable internships. Educational opportunities offered by various internships and residencies differed greatly; in many hospitals they offered only limited experience and responsibility. Medical school graduates often selected a hospital for its abundance of patients and its wide variety of diseases. Here city hospitals had the advantage—an intern at Los Angeles General Hospital reported that his internship had given him the opportunity "to watch and care for 1,000 patients"—and in the 1930s and 1940s applicants were thronging to them. At Bellevue a superintendent is reported to have told an intern who had asked him to censure an orderly: "I'm sorry, doctor, but there's nothing I can do. We have hundreds of interns panting at the gate, but a good orderly is hard to come by."[74]

Despite long hours and hard work, few house officers quit once they had started. For one reason, they were self-selected: those who were able, energetic, ambitious, and skillful at organizing their activities were most likely to apply to city hospitals. Second, once they began this frenzied Indian war dance of life against death, the conscience of the corps took over. They dared not desert their patients and their colleagues; they became the iron men who could endure all hardships, and they boasted about it ever after.[75]

Under these circumstances house officers were targets for exploitation. In industry an excess of labor means low wages. In city hospitals it meant practically no wages and hard, unremitting work. As late as 1952 interns received board and room plus $10 a month at Baltimore City Hospitals, $15 at Charity (New Orleans), $50 at Gallinger Municipal (Washington), $55 at St. Louis City, $64 at Cleveland City, and $92 at Boston City Hospital. Thus, the young doctor's zeal to learn and his professional motivation to do the job right could be manipulated by politicians whose main object was to win votes. A symbol, almost a caricature, of such exploitation was Mayor Curley, strutting through the corridors of the Boston City Hospital with a fat cigar in his mouth, patting every intern he met on the back and telling him, "You fellows are doing a f-i-n-e job."[76] Such exploitation was facilitated by spineless executives who were afraid to tell their superiors and the public what would be needed to provide acceptable medical care. The system of making the house officer the point man of the squad, a victim of the mistakes and omissions of those above and below him, did not change materially until house officers throughout the United States rebelled in the 1960s (see Chapter 8).

Undergraduate medical education in city hospitals improved after 1910 even more than did house officerships. These hospitals began allowing medical students to act as members of the team of attending physicians and house officers when university services were inaugurated. At the Boston City Hospital this occurred by 1923 or shortly before, and model clinical clerkships had been developed by 1926 on the Cornell Medical Division at Bellevue.[77] In Baltimore clinical clerkships became attractive to the medical schools in the 1930s, when the hospital was enlarged and modernized, and at about the same time genuine clinical clerkships were provided for Georgetown and George Washington medical students at Gallinger Hospital.

At Cook County Hospital an attending physician had to get special permission to take students onto the wards as late as 1926.[78] Gradually, in succeeding years the students' presence there came to be tolerated, but without organized services the medical schools could not give the supervision needed for proper clinical clerkships. Finally, in the 1950s the professors of medicine in the four medical schools teaching at the hospital developed clinical clerkships for students on services staffed by their faculty members.[79] The varied experiences of medical schools in these four city hospitals typified arrangements made in others.

As for the general medical profession, the majority of practicing doctors paid little attention to the city hospitals. Preoccupied with their own practices and busy trying to keep up with technical advances in medicine, they tended to assume that the doctors who spent a large part of their time at the city hospitals would see to it that necessary administrative as well as strictly medical matters would be taken care of. I encountered this attitude among Baltimore doctors with respect to Boggs and Shipley at the city hospital, among Chicago doctors with regard to Meyer at Cook County, and among Washington doctors in relation to Hussey and me at Gallinger Hospital. Occasionally practicing doctors would become uneasy because they thought a city hospital was caring for patients who otherwise would have come to them, as in Baltimore in 1960 (see Chapter 7).[80] Aside from infrequent complaints such as this and an occasional report made by a committee at the request of a medical society, the general medical profession usually left the city hospitals to face their own problems.

During the first half of the twentieth century many city hospitals and medical schools moved from a loose association to highly organized services run by the schools within the framework of hospital policy. At first, individual doctors controlled particular wards, house officers fit in where they could, and students stood on the periphery. Later, doctors selected by medical schools on the basis of medical knowledge, teaching and research abilities, and leadership qualities directed a unified team of staff members, house officers, and students. Where these programs were encouraged and supported, medical education, research, and the training of house officers reached a high level of excellence, and medical care was superior.

# 7 · Medical Care: Running to Keep Up

> It is like a nursing mother to the city's
> poor...The poorest receive from the same
> hands what the rich pay heavily for.
> >                  Description of Cook County Hospital,
> >                                                1941

> We have accepted it...that the patients
> who come to the city hospitals are just not
> worth as much as those who use the pri-
> vate hospitals and pay for it.
> >                                    Lewis Thomas

IN 1910 AMERICANS were entering a half century of phenomenal improvement in health. A child born in 1960 could be expected to live twenty years longer on the average than one born in 1910. Deaths from infectious diseases such as tuberculosis, typhoid fever, diphtheria, measles, and intestinal infections fell sharply. While preventive measures, a better standard of living, and perhaps increased resistance to infecting microorganisms accounted for part of the decline, specific serums and drugs were major factors. The spectacular fall in the number of deaths from appendicitis reflected the combined effects of modern surgery and antibiotic therapy. After a gradual rise from approximately seven deaths per one hundred thousand population in 1900 to a peak of fifteen in 1930 (probably as a result of more exact diagnosis), the mortality rate fell sharply, reaching one per one hundred thousand in 1960.[1] Since nearly all patients with appendicitis reached the hospital, this improvement was a good example of the effect of better hospital care. Patients with many other illnesses who were hospitalized in the 1960s were either restored to health more quickly and completely or made more comfortable than in 1910.

But new drugs and serums and supplies and equipment for better diagnosis and treatment cost money—lots of money. In the past the city hospital's reach had always exceeded its grasp; its aspirations for its patients had always been beyond its resources. Now the gap between the ideal and the actual was greater than ever. No city hospital could afford the optimal in every area; most of them were deficient in several or many areas. In a few the buildings were so old, dilapidated, and poorly kept, the wards so crowded, unkempt, and disorganized, the drugs and supplies so scantily provided, the equipment so deficient, outmoded, and outworn, and the doctors, nurses, and attendants so overworked that at times a patient's survival could be attributed less to good nursing and the use of the new remedies than to the recuperative powers of the human body and the strength of the human spirit.[2]

Most patients' initial contact with a city hospital was through the admitting clinic. It was a painful beginning. A committee investigating Cook County Hospital in 1936 reported that patients and their families filled all the seats of the barnlike waiting room, and many were forced to stand for hours. All along a nearby corridor patients on stretchers endured the same interminable wait, terrified and alone. Examined at last by a hurried, overworked, abrupt intern, often one with little experience, a patient ill enough to be admitted suffered through another long wait until an orderly or an intern was found to take him on a wobbly stretcher through endless passages to a drab, dismal ward. An investigating committee described these practices as "grossly crude, approaching the inhumane"; yet they were similar to conditions in admitting clinics of many city hospitals.[3]

On the wards the patients were little better off. A surgeon who interned at Bellevue in 1953 recalled a typical surgical ward, planned for sixteen patients but usually containing thirty-two and providing no screens or partitions for privacy. There were never enough nurses; on late shifts only one nurse cared for all thirty-two patients. She did the best she could for the sickest patients, who had "intravenous solutions running into their arms, gastric suction tubes running out of their noses and catheters running out of their bladders. There were usually one or two vomiting and a couple of others moaning in pain or yelling for their hypos. . . The not-so-sick patients at the far end more or less shifted for themselves."[4]

The half century following 1910 witnessed a continuous struggle to provide proper nursing for city hospital patients. The establishment of departments of nursing under competent nurse-administrators and the founding of hospital nursing schools had been great steps forward, but many problems remained. Nurses were expected to work long hours for low pay and to accomplish more than was humanly possible in those hours. City hospitals, with their lower prestige, unattractive physical surroundings, and smaller outlay per patient (see Table 10, discussed later in this chapter, for comparison of costs in public and private hospitals), found it hard to recruit and retain enough nurses. The small nucleus of devoted nurses who kept the nursing services going, no matter how able, resourceful, resilient, and seemingly tireless, was never enough to do the job as it needed to be done.

Cook County Hospital was especially deficient. Experts surveying the hospital in 1927, though impressed with the highly efficient nurses, stressed that they barely sufficed for the number of patients. The ratio of graduate and student nurses to the number of beds was 1:6.1 (see Table 9). This ratio was much better in ten other city hospitals, varying from 1:2 at the Buffalo City Hospital to 1:4 at Bellevue.[5] The effectiveness of the few nurses on duty was lessened by the physical conditions under which they had to work (described by an expert as the worst in any hospital of its type in the United States, involving overcrowding, long distances from nurses' stations to patients, and lack of a call system and other essential equipment.) Living arrangements for nurses and students were also poor, but this deficiency was remedied when a handsome nurses' home and school was completed in the mid-thirties. Meanwhile, salaries remained low, recruiting was unproductive, and poor physical conditions still hindered proper nursing.[6]

During the Depression there was an oversupply of nurses nationally, and the hospital rocked along without changing its policies.[7] But as the demand for nurses increased with the rapid specialization of medicine, and as the supply was threatened by the opening of other more attractive vocations for women, hospitals like Cook County felt the pinch. In 1950 outside consultants warned that standards of nursing care were so low as to be dangerous. The hospital had only one-third as many persons caring for its patients as deemed necessary by the National League of Nursing (1,192 compared to 3,341). When only registered nurses were considered, the discrepancy was even greater: 383 versus 2,219, or about one-sixth of the generally

**Table 9.** Ratio of nurses and attendants to beds in selected public-general hospitals, 1927.

| Hospital | Capacity (beds) | Ratio of personnel to total capacity | | | | |
|---|---|---|---|---|---|---|
| | | Graduate nurses | Student nurses | All nurses | Attendants | Total nurses and attendants |
| Ancker, St. Paul | 850 | 1:37 | 1:3.8 | 1:3.5 | 1:44.7 | 1:3.2 |
| Bellevue, New York City | 1733 | 1:7.9 | 1:8.7 | 1:4 | 1:3.3 | 1:1.9 |
| Buffalo City | 401 | 1:5.6 | 1:3 | 1:2 | 1:19 | 1:1.8 |
| Cincinnati General | 800 | 1:10.8 | 1:5 | 1:3.5 | 1:10.5 | 1:2.6 |
| Cook County, Chicago | 1600 | 1:17 | 1:9.4 | 1:6.1 | 1:5.3 | 1:2.8 |
| Detroit Receiving | 365 | 1:3.6 | 1:9.9 | 1:2.7 | 1:7.3 | 1:1.9 |
| City, Jersey City | 600 | 1:11 | 1:5 | 1:3.5 | 1:8.5 | 1:2.5 |
| Los Angeles County | 1283 | 1:5.3 | 1:4 | 1:2.4 | 1:6.9 | 1:1.8 |
| Milwaukee County | 308 | 1:15 | 1:4 | 1:3 | 1:13 | 1:2.4 |
| Minneapolis General | 634 | 1:13.5 | 1:4.8 | 1:3.5 | — | — |
| Philadephia General | 1600 | 1:17 | 1:4.2 | 1:3.4 | 1:5.3 | 1:2 |

Source: Figures adapted from M.T. MacEachern and E.W. Williamson, "Report of Survey: Cook County Hospital, Chicago, Illinois" (mimeographed, Chicago: Hospital Standardization Department, American College of Surgeons, 1927), pp. 5–6, 7.

accepted minimal requirement. Fourteen wards containing 991 patients did not have a nurse on duty at night; twenty-six briefly trained attendants supplied these patients with whatever nursing they received. Two supervisors, the only nurses on duty on these wards, gave whatever emergency medications they could find time for. As a consequence, patients lay on sheets soiled with urine and feces for up to six hours, day or night. Helpless ones might be fed by other patients or be given cold food if an attendant got around to them; often they went without. No medicines or treatments were given on medical wards during the night shift unless a serious emergency occurred and a supervisor was free to go to the ward. House officers stated that they went to these wards late at night to administer narcotics because they could not sleep knowing a patient was suffering there. The scarcity of registered and licensed practical nurses shifted most of the patient care into the hands of nurses' aides, who had only three weeks' training and often were barely literate. Doctors were forced to depend on them to detect changes in a patient's physical or emotional condition.[8] As shown in Table 9, Cook County Hospital had more attendants per patient than many other city hospitals.

Such pitifully poor nursing care could be explained simply: hard work, poor facilities, and low pay sent nurses elsewhere, and the added vacancies increased the work for those who remained. Only $2.35 per patient day was spent for nursing care at Cook County Hospital in 1949, compared with an average of $4.63 at eight other city hospitals and $4.51 in seven state-owned general hospitals operated as university hospitals. When another committee surveyed the hospital in 1963, nursing care had improved little if at all. The hospital continued to fail miserably in its task of caring for the sick poor of Chicago.[9]

Other city hospitals faced similar dilemmas: increasing demands upon nurses, along with public apathy and inadequate budgets. A committee of trustees and medical staff of the Boston City Hospital reported in 1916 that the average number of beds per nurse varied from 1.5 to 2.6 in five private or university hospitals, while at Bellevue and Boston City Hospitals there were over four beds per nurse. At the Baltimore City Hospitals the chiefs of medicine and surgery pointed out in 1926 that the burden of providing satisfactory nursing could not "be shifted indefinitely on a few devoted people." "First and last," they contended, "the solution of our difficulties lies in adequate personnel, adequately paid."[10] But when the superintendent asked for 383 additional employees in 1933, the chairman of the board of supervisors opposed the request because raising the money would require a 6 percent increase in the tax rate. Fifteen years later a citizens' group declared the employee-patient ratio at the city hospital to be substandard and well below the average for similar institutions. And so matters continued through the years. As late as 1970 the chairman of the governing board informed the mayor that only thirty-eight of sixty-six nurses' positions were filled, because other Baltimore hospitals paid nurses substantially more than the city hospital.[11] The more things changed, the more they remained the same.

City hospitals across the nation tried to make up the deficit in nursing services by means of nursing schools. These schools had been welcomed by hospital authorities, beginning in the late nineteenth century, not only because they produced better nurses than the former hit and miss training but because students cost the hospital less than trained nurses and performed many of the same tasks (see Table 9 for figures on the use of student nurses on the wards of city hospitals). But this situation gradually changed. Advances in scientific medicine demanded more of nurses, and newly organized national nurses' associations set out to see that graduates of the schools measured up to the demands.[12] By 1914 forty states had laws listing requirements for registered nurses. Rising standards of instruction elevated costs, and when it became cheaper to staff hospitals with graduate nurses, many hospitals closed their schools. The number of diploma (hospital) nursing schools in the United States fell from about 2,150 in 1926 to 1,300 in 1946. Despite a temporary reversal during the Second World War, the decline continued in the face of an increase in the number of patients admitted to hospitals.[13]

Replacing the hospital nursing schools were the collegiate schools, which grew slowly around the turn of the century and then more rapidly beginning in the thirties. In the period from 1953 to 1976, when nationwide figures are available, graduations from diploma programs decreased by one-fourth (from 26,824 to 19,861), while bachelor's degrees granted in nursing rose tenfold (from 2,171 to 22,678) and the number of associate degrees jumped astoundingly from 260 to 35,094.[14] Because city hospitals were in less accessible and sometimes even dangerous neighborhoods, because the surroundings were less attractive, the wages lower, and the impediments to good nursing greater, their nursing schools were hardest hit by the rise of collegiate nursing programs. The history of the school at Washington's city hospital illustrates this change. Like Cook County and Baltimore City, this hospital had retained a small nucleus of loyal, hard-working nurses despite perennially low budgets. But there were never enough. At the Capital City School of Nursing, as in other schools, the superintendent of nurses, ward nurses, and doctors had been the early teachers. In 1923, after the student body had reached twenty-seven, the first full-time instructor was added. Until facilities at the Washington Asylum Hospital were complete, nursing students had to spend up to one-third of their three years elsewhere, mainly for experience with children. In 1930, after a pediatric service had been established, the last tie with other hospitals was loosened, and the school was on its own at last. By then fifty-seven students were enrolled.[15]

Students came from middle-class homes, many from small towns or rural areas. They were expected to work hard and did. Some idea of the rigors of their training can be gathered from a letter written by the president of their alumnae in 1936:

> We were expected to work from 7 to 7, six days a week, and from 7 until 12 on every other seventh day...
> This work is actual physical labor, and...it permits no time for rest...some of this work is mental torment, that allows no escape...much of this work...is sickening and filthy;...It is dangerous work. It turned one of three of us from fulfillment of those ideals which made the rest of us nurses; it contributed to the tuberculosis which has stolen years of youth from one of every ten.

The writer then listed the impediments to good nursing—lack of space and privacy for patients, lack of attendants, lack of enough food for patients and nurses—and warned that the nurses' examining board of the District of Columbia might no longer accept candidates from the school unless deficiencies were corrected.[16]

Student nurses were exploited not only with overwork but by being given responsibilities beyond their experience. A nurse who trained at Washington's city hospital in 1936 wrote: "As soon as I received my cap [finished a few months' probationary training] I was placed on night duty [in charge of a ward] alone." Although outside experts observed "whole-hearted zeal in the care of patients" and "no indifference, frivolities, or distractions from the business at hand, nor any evidence of defective morale," Congress continued to appropriate just enough funds to cover the worst deficiencies and placate certifying bodies. Shortages of personnal, equipment, and supplies continued, and the warning of rejection by the nurses' examining board was repeated many times.[17] Despite these handicaps, unbiased observers in the late 1950s still found the care of patients to be "considerate, sensitive and kind." And through all vicissitudes the alumnae remained loyal. In 1950, for instance, they paid the fee required to have their school inspected for national accreditation. It passed.[18] But except for the hospital staff, the alumnae, and a small group of society women, few knew about and fewer appreciated the Washington school's achievements. Eventually the handicaps became too great; after two-thirds of a century of honorable service against great odds, the school gave up. The last class graduated in 1972.[19]

Cook County's nursing school strove hard to retain its reputation as a leading school. At midcentury it was classified in the upper 25 percent of nursing schools in the nation. The Second World War had produced a peak enrollment of 140 students, which was only two-fifths of the school's capacity. After the war enrollment fell to twenty-nine in 1947 and had risen only to fifty-nine by 1949. Moreover, in 1944 54 percent of the students withdrew from the school during their first year, 49 percent in 1948. Outside consultants found the load of classroom, laboratory, and ward work in the first year to be excessive. It was as if the school's leaders were demanding that the students themselves, by superhuman effort, make up for the school's defects: insufficient and sometimes inexperienced teachers, inadequate ward facilities and supervision, difficult nursing conditions, and dispiriting surroundings. The administrators, teachers, and students deserved great credit for their dedication. The nurses who managed to graduate had profited from experience in a thousand trials, but their education had been obtained at great cost in human lives—theirs and others'.[20]

In addition to poor funding and public indifference, Cook County's nursing school was further hampered by its remoteness from the hospital's center of control. The nonprofit corporation that ran the school was responsible, through a contract with the Board of County Commissioners, for furnishing nursing care for all patients in the hospital and for the social service department and some dietary services. Yet, despite the brave facade of an independent corporation and the prominent names on its board of trustees, the school possessed no property of its own and depended wholly on the county commissioners for its yearly budget, approximately $4 million in 1950. By closing their eyes to much that went on around them, the school's leaders achieved a singleness of purpose that helped them train good nurses in the midst of chaos. While remaining aloof from the hurly-burly of the hospital's operations, the school could not hope to modify the hospital's policies; instead it was the victim of those policies. Thus, both school and hospital suffered because the nursing school had so little influence. Outside consultants, noting that the school had not succeeded in attracting appreciable amounts of money as gifts in recent years or kept the nursing service or school out of politics, labeled the separate corporation cumbersome and inefficient and recommended that nursing functions be subordinated under the hospital administration. But dual authority was not abolished until 1969, when both hospital and nursing school were placed under a newly created Health and Hospitals Governing Commission, and a decade later the decision was made to close the school (see Chapter 8).[21]

While these two nursing schools added much to their respective city hospitals—fresh recruits, dedicated nursing services, and sorely needed administrative machinery at the ward level—their contribution was limited by deficiencies in hospital budgets and physical plants. The nursing service at the Boston City Hospital, facing fewer of the handicaps of other city hospitals, maintained a little longer its tradition of a strong nursing school. It was highly popular among Irish Catholics and drew recruits who were superior in intelligence and motivation. Discipline was strict: "Nurses did not talk up;...the old ladies ran things with an iron hand." Yet training at the hospital stimulated nurses to thrive on challenges and aspire to positions of leadership. Relations with doctors were good; each group respected the other, and they worked together by mutual agreement. Ward nurses were the hospital's middle administrators, overseeing housekeeping, ordering supplies, keeping the ancillary help in line.[22] The school also contributed to the hospital by attracting funds from private donors (see, for instance, Chapter 4).

Beginning around midcentury, as the increasing application of scientific medicine to patients introduced large numbers of technicians and professionals into the hospital and more complex organization brought more administrative and clerical personnel, nurses in city hospitals became increasingly restricted to their own affairs. By the 1960s the combined effects of old, outmoded hospital buildings, overcrowded, out-of-date nurses' residences, and growing competition from collegiate degree programs were taking their toll on this school also. In 1967 there were twenty-two vacancies in an entering class of one hundred, compared with fifteen out of 135 in the diploma program of the Massachusetts General Hospital and fifteen out of 120 in the degree program of Northeastern University.[23] Although physical improvements and new leadership improved the image of the school, the realities of declining appropriations finally forced the hospital to give up its nursing school. The last class graduated in 1975 (see Chapter 8).

In contrast to these hospitals, the Baltimore City Hospitals kept its sights lower. The nursing school that opened in 1925 became a one-year school for practical nurses because hospital authorities never succeeded in obtaining funds for enough instructors to satisfy accrediting agencies. Within these self-imposed limits the school produced good nurses who performed well under the supervision of registered nurses; yet competition from academic programs likewise forced it to close in 1975.[24]

Thus, these four nursing schools, with differing programs, organizations, and degrees of support, were all compelled to close eventually. With the demise of the hospital nursing schools went a major strength of the city hospitals. Besides providing cheap help and a pool for recruiting graduate nurses, the schools' staffs formed a stable organization through which many of the functions of the hospital could be carried out. At times, in certain hospitals, dietary, social service, and housekeeping operations had been placed under the departments of nursing. Even when they were not in control, nursing departments influenced the actions of administrators, doctors, and others. By maintaining connections with civic-minded citizens, and particularly with women because of a mutual interest in women's rights, they were able to get the city hospital's case presented in high places and attract funds needed to supplement the meager budget.[25] Despite their deficiencies, nursing schools in city hospitals furnished excellent opportunities for nurses in training; they provided wide experience in a variety of diseases and stressed practical measures. In 1949 six national nursing organizations rated hospital nursing programs, both public and private, in two groups according to quality, listing the top 25 percent and the middle 50 percent of all programs evaluated. (Programs in the lowest 25 percent were not listed.) Of the 291 programs in the upper 25 percent, twenty-three were in large city hospitals, whereas only five of the 468 programs in the middle 50 percent were in large city hospitals, a highly significant difference (p = $<0.001$). Nursing schools in the top 25 percent included those at Bellevue, Boston City, Charity (New Orleans), Cook County, and Los Angeles County Hospitals. Among the pioneer schools at city hospitals, only the Philadelphia General fell in the middle half.[26] Thus, while every hospital lost something of value when its nursing school closed, city hospitals were generally hurt more than others.

Other factors besides nursing vitally affected patient care. Physical surroundings on the wards of city hospitals frequently fell below the standards for healthful living, mimicking the environment in the poorest tenements. In 1938 a Chicago citizens' committee listed conditions at Cook County Hospital that would have produced an uproar if found in a private hospital: unworkable sterilizers, showers, refrigerators, and drinking fountains; broken, hard-to-manage stretchers and wheelchairs; a shortage of bedpans, bed screens, instruments for applying surgical dressings, and patients' gowns, bedclothes, and towels. The absence of a call system forced patients to scream or pound bedside tables to summon a nurse, and, if too weak to do so, perhaps to die. There were no facilities for taking roetgenograms in the tuberculosis or the men's medical building (each containing several hundred patients); when an X-ray examination was needed, the patient had to be transported several blocks, often on a broken stretcher by a reluctant orderly—if one could be found.[27]

Most of the $35.9 million obtained for the Cook County Hospital from four bond issues over the next three decades was spent on repairs, renovations, and major equipment. Yet, as late as 1967 another citizens' committee reported that shortages of toilets and bathing facilities practically prohibited proper personal hygiene, and the large, open, usually crowded wards and the paucity of screens afforded little opportunity for privacy or personal dignity.[28] The chief of a medical division complained that often whole wards were out of essentials ranging from linens to penicillin and fluids for intravenous administration. Interns were still forced to run from floor to floor to find enough tubes for blood testing.[29] Cook County Hospital may have run hard to catch up, as its administrators claimed, but it was still behind many city hospitals and far behind most private hospitals in patient care.

At the city hospitals in Washington and Baltimore, after modern buildings were opened in 1929 and 1935, respectively, supplies and equipment could be characterized as skimpy but usually manageable. Since budgets were tight in both hospitals, supplies were dealt out parsimoniously, and long waits for new equipment were common. Generally, these inconveniences and delays were not severe enough to undermine morale or seriously hamper patient care. Yet, in 1967 a congressional subcommittee described laboratory equipment at the District of Columbia General Hospital as "so insufficient and archaic" that "complete accuracy of results" was impossible.[30]

In the Boston City Hospital during the free-spending days of Curley, housekeeping was better than in most city hospitals (especially because Curley's expanded patronage system provided three helpers where one would have been sufficient), equipment was usually up-to-date, instruments and supplies were fairly adequate, although at times red tape in City Hall might slow the arrival of a new shipment for weeks or months.[31] As the city started to feel the pinch of overspending in the 1950s, the hospital began to take on the seedy look of other city hospitals, and administrators scrimped on supplies and equipment. By 1958 a reporter described the hospital as "mostly shabby," while giving it high marks for patient care and staff morale. She called most of its fifteen hundred beds obsolete and related how the shortage of incubators forced the staff to keep premature babies in baskets with hot bricks or electric irons to warm them. But deficiencies were mostly episodic and spotty, and the staff took them in stride. The superintendent expressed the spirit of the hospital: "Only one kind of a general hospital can be operated in this city and that is a good one."[32]

Since scarcity and want prohibited ideal patient care, it is not surprising that city hospitals were vulnerable to criticism. Yet, denunciations appeared less often in the medical or lay press than might have been expected, considering that the city hospital usually cared for more patients than any other. When the news media did take a city hospital to task, however, its already tarnished reputation was further blackened, especially when headlines screamed "Misery Harbor" (Cook County Hospital), "The Butcher Shop" (Lincoln Hospital, New York City), "Our Murder Factory" (Gallinger Municipal Hospital).[33] Hospital authorities usually reacted by denying accusations where possible rather than using them to obtain higher budgets. Cook County Hospital officials believed that if anyone dared to censure them, offense was the best defense. In 1963, when the Joint Commission on Accreditation of Hospitals listed sixteen shortcomings that placed the hospital's accreditation in jeopardy, Karl Meyer brushed them aside, claiming the commission was making the hospital cross *t*'s and dot *i*'s and telling a reporter, "We can't meet all this tommyrot." Yet, two years later the county commissioners finally admitted something was wrong when they appointed an administrator to take overall authority at the hospital, although not in the nursing school.[34]

In contrast, when Baltimore's Commission of Governmental Efficiency and Economy criticized a number of features of the city hospital in 1949, the director of the Department of Public Welfare sent a temperate letter to the mayor pointing out that the deficiencies cited were attributable to insufficient supervisory positions and low salaries and alerting the mayor that remedies would be recommended to him promptly. The welfare director stressed the excellent work done at the hospital, referring to a recent survey by the American College of Surgeons giving the hospital a high rating for professional services, and concluded that there was nothing wrong with the hospital that could not be remedied if the mayor provided the kind of support that he had always promised. Likewise, in 1966, when Baltimore's controller criticized the city hospital, the City Hospitals Commission expressed hope that the report would at least help obtain much-needed funds for personnel and equipment.[35]

Perhaps most officials did not respond constructively to criticism because they were battle weary. A District of Columbia commissioner told a Senate subcommittee investigating Gallinger Hospital: "I know what these investigations are . . . They [Congresmen] get an investigation in July and recommend something, and then the next July we [the commissioners] go to the Appropriations Committee and we don't get what we ask for."[36] In the nation's capital differences among the various interests, present in every city, were more readily apparent when focused through the lens of congressional hearings: doctors, nurses, and patients pleaded for more manpower, better equipment, and adequate supplies; congressional committees represented the indifference of the ruling classes and the bulk of the taxpayers; city and hospital authorities were caught in the middle.

De Hartog attributes the infrequency of protests by city hospital staffs to what he calls the disaster syndrome, "when normal reactions of protest or outrage in the face of intolerable conditions are absent and the overriding reaction is not to correct those conditions but to accept them as permanent and to circumvent them."[37] I have been informed that the syndrome operated frequently at Cook County Hospital, where interns diligently searched for and carefully hid the few tubes available for collecting blood so as to make sure that *their* patients' blood would be tested.

Crowding was a problem in all city hospitals. At Cook County Hospital it was continuous throughout the period, if anything becoming worse. In 1935 the capacity was 2,600, but a daily average of 2,619 patients was maintained by placing beds in hallways and opening two new floors in the children's building. Despite these maneuvers, 82,721 patients were turned away from the hospital. As late as 1965 a staff report condemned the nursery for premature babies as overcrowded, ill-ventilated, and inadequately equipped, and two years later one of the succession of committees to inspect the hospital stated flatly that poor facilities made quality patient care impossible. The large, open wards, relics of turn-of-the-century hospital architecture, had been designed for two rows of beds, but two additional rows had been packed in the center area for so long that this was accepted as standard practice.[38]

In the city hospitals of Boston and Washington crowding was episodic. Winter months usually saw beds in aisles and occasionally in hallways. Sometimes new buildings would be erected to provide room for certain services, but usually only after requests for relief had been refused for years or decades, during which time more and more patients were packed into existing facilities. Then space in some units would be adequate for a few years, whereupon the whole cycle would begin again.[39] The aggressive building program of the Boston City Hospital caused some large wards to be replaced with rooms and small wards as early as the 1930s. The less vigorous building program of Washington's city hospital produced similar changes, although fewer and more slowly.

At the Baltimore City Hospitals overcrowding of wards in the general hospital was seldom a problem after an efficient administrator was installed in 1924. One reason for this was that dumping, or transferring undesirable patients to a city hospital by other hospitals without advance arrangement, was rare in Baltimore because the sending hospital had to obtain permission from the receiving hospital before a city ambulance could transport a patient. Furthermore, the city paid for the care of a large number of patients in private hospitals. In 1930 the city hospital cared for only 25 percent of Baltimore's indigent patients (see Chapter 5); in 1937, after the new general hospital building had been opened, for 40 percent; and in 1956–1957, 36 percent. By contrast, in Washington in 1930, after the first modern general hospital building had been in operation for a year, 81 percent of the patients needing the city's financial assistance to obtain hospital care were hospitalized at Gallinger Hospital. In 1961 this proportion was still 81 percent. In New York City in the 1960s about 60 percent of the hospital care of the indigent sick was given in municipal hospitals.[40]

Although authorities denied it and the public ignored it, the plain truth was that cities did not give their public hospitals enough money to run properly. Proof could be found in comparisons with private hospitals. In 1906 Goldwater wrote that expenditures per patient per day were twice as great at private as at city hospitals (see also Table 5), and this ratio continued until midcentury or later. Data collected in 1925 for Boston, New York, and Chicago (Table 10) demonstrate this. For instance, the average expenditure per patient per day in New York City in four leading private hospitals was $6.06, compared with $3.26 in four city hospitals. In Washington in 1938 the discrepancy was even greater. The city paid from $6.29 to $6.72 per patient day

Table 10. Comparative costs for patients in selected public and private general hospitals, 1925.

| City | Public hospitals | | Private hospitals | |
|------|------|------|------|------|
| | Name | Cost/patient/ day | Name | Cost/patient/ day[a] |
| Boston | Boston City | $4.30 | Massachusetts General | $6.83 |
| | | | Peter Bent Brigham | 6.04 |
| Chicago | Cook County | 2.99 | Presbyterian | 7.40 |
| | | | Wesley | 8.67 |
| New York | Bellevue | 3.36 | New York | 6.44 |
| | Fordham | 2.95 | Presbyterian | 7.39* |
| | Gouverneur | 3.72 | Roosevelt | 5.61 |
| | Harlem | 3.00 | St. Luke's | 4.80* |
| | Average, four hospitals | 3.26 | Average, four hospitals | 6.06 |
| Average | Six public hospitals | 3.39 | Eight private hospitals | 6.64 |

Sources: "Average Daily Census and Per Capita Cost for 60 Hospitals As Shown by Reports," Hospital Management 23 (June 1927):51; Cook County Hospital, Annual Report, 1925 (Chicago: James T. Igoe, 1925), p. 70; Trustees of the Boston City Hospital, Annual Report, 1925 (Boston, 1926), p. 26.

a. All figures are for private and ward patients except those marked *, which are for ward patients only.

for patients at three private hospitals, compared with $2.39 per patient day at Gallinger. (Costs at Freedmen's, a federally subsidized hospital, were between the two, $3.98 per day).[41] Because of these extra dollars, the private hospitals generally provided all their patients—those in wards as well as those in private rooms—with cleaner, less cramped, more attractive quarters, better food, and more nursing care than city hospitals. Thus, medical care in city hospitals was always a problem trying to do too much with too little. Furthermore, some city hospitals lagged far behind others in expenditures (see Table 11). In 1925 Boston spent more than Washington and Cook County and more than twice as much as Baltimore for each patient at its city hospital. By midcentury expenditures at Boston's hospital were slightly above the average for all short-term hospitals in the United States, at Baltimore and Washington distinctly below, and at Cook County far below the average. By 1965 all were above or near the national average.

Since costs were ultimately determined by the amounts budgeted by the controlling board, and these were severely restricted, hospital authorities sought outside sources of support. One way to increase the

Table 11. **Expenditures per patient per day in four public general hospitals, 1925–1965 (in dollars).**

| Year | Baltimore City Hospitals[a] | Boston City Hospital[b] | Cook County Hospital | Gallinger Municipal— D.C. General Hospital | Average, all short-term U.S. hospitals |
|------|------|------|------|------|------|
| 1925 | 1.65 | 4.30 | 2.99 | 3.60 | — |
| 1935 | 2.74 | 5.33 | —[c] | 2.09 | — |
| 1950 | 12.48 | 16.59 | 4.82[d] | 12.50 | 15.62 |
| 1960 | 31.02 | 46.88[e] | 11.67[d] | 29.00 | 32.23 |
| 1965 | 49.42[f] | 57.62[e] | 38.72[d] | 43.46 | 44.48 |

*Sources:* Annual reports for the individual hospitals; for Boston in 1960, see *Hospitals,* Guide Issue 35 (August 1, 1961):104; for Boston and Chicago in 1965, see ibid. 40 (August 1, 1966):69, 106; for average of all short-term hospitals, see U.S. Department of Commerce, Bureau of the Census, *Historical Statistics of the United States: Colonial Times to 1970.* 2 vols. (Washington, D.C.: U.S. Government Printing Office, 1975), 1:81.

a. General hospital for short-term care.

b. Hospital proper (not including South Department).

c. Annual reports during this period give no financial data.

d. Calculated from average census and total expense figures; includes psychopathic hospital.

e. Calculated from average census and total expense figures; includes East Boston Relief Station.

f. Figures from last annual report available (1964); American Hospital Association figures include chronic diseases hospital.

income of municipal hospitals and at the same time raise their prestige was to admit patients who could pay some or all of their hospital expenses. Boston City Hospital had followed this policy from the beginning by providing rooms suitable for private patients (see Chapter 4) and charging moderate fees to those able to pay.[42] But for many years most city hospitals saw their job as dispensing charity to those who could afford nothing better. At the prodding of legislators, the hospitals made some attempts to collect from patients admitted as emergencies who were able to pay, but these efforts were often ineffectual, in part because these same legislators refused to provide the staff necessary to do the collecting.

The picture changed with the rise of Blue Cross insurance in the 1930s and 1940s and the passage of the Social Security Act in 1935, which granted assistance to states for the care of the aged, the blind, and mothers of dependent children. Subsequent amendments and other federal statutes enlarged these benefits, while private sickness insurance plans began to grow. A giant step was the provision of medical care for persons over the age of sixty-five through the Social Security Amendments of 1965. In the thirty years from 1939 to 1969 the number of persons covered by hospitalization insurance increased from fewer than eight million to over 170 million (from 6 to 85 percent of the population).[43]

For a long time city hospitals did little to adjust to this profound change. Baltimore's city hospital was a good example. The 1938 report of the Department of Public Welfare contains a muted statement: "If the patient or his relatives are able to defray part of the expenses of hospitalization the attempt is made by the admitting office to obtain a promise of such partial reimbursement." The machinery for collection was, according to the report, being "perfected." In view of this hesitant approach, it is not surprising that collections were scanty. Not until 1951 was a ruling obtained from the city's legal department that patients with hospital insurance could be forced to pay charges assessed by the hospital. As late as 1955 the superintendent maintained that the hospital was intended only for indigent and semi-indigent patients—in other words, those who were already flat broke or being flattened by their illness. (The semi-indigent are also described as medically indigent; they can pay the ordinary expenses of living but cannot pay their way through a major bout of illness.) A single person with a minimum income of less than $1,000 annually was not expected to pay anything, and one with an income between $1,000 and $2,500 was expected to pay part of his hospital expenses. Single persons with incomes above $2,500 were supposed to go elsewhere; the staff was instructed not to admit them except for treatment of emergency illnesses and communicable diseases.[44]

Baltimore's city hospital had been following this prudent policy to minimize complaints from private physicians that the hospital was infringing upon their practices; but despite this precaution the doctors' simmering resentment boiled over three years later. By this time the medical staff had achieved university-type services, composed of both full- and part-time doctors dedicated to developing high-grade medical care. The senior staff proposed admitting a few paying patients to diversify the illnesses encountered and thus improve the training of house officers. The Department of Public Welfare approved, but the city medical society objected violently. Medical staff and hospital authorities persisted, and their efforts eventually culminated in the formation of a group practice organized by the medical staff, which contracted to supply complete medical care to private patients in areas near the hospital (see Chapter 8). When these patients needed hospitalization, they were admitted to the city hospital, and the hospital was reimbursed through insurance. This arrangement helped banish the poorhouse image and assured the hospital of a steady income in addition to city appropriations and a constituency of citizens who supported the institution politically. It attracted high-caliber doctors by assuring them a good income, and it provided better training for house officers, while the patients benefitted from improved morale and a superior staff.[45]

The arrangement was hailed as the answer to many of the problems of city hospitals.[46] Yet, it should be noted that Baltimore's city hospital was aided in this venture by special circumstances: it was situated in a middle-class, blue-collar neighborhood rather than in a severely deprived area as in Boston or Chicago and was closely affiliated with a medical school. The new plan was in line with Baltimore's tradition of working with the academic community to solve health problems rather than leaving them to politicians, as in Chicago, or approaching each problem as a separate crisis and trying to solve it according to the prevailing theory or the power in ascendancy at the moment, as in Washington. Perhaps this principle of longer-range planning was more significant than the specific plan adopted. At any rate, the response of the Baltimore City Hospitals to the radical changes in the economics of health care was imaginative and forward looking and may well be a harbinger of the future.

In the nation's capital the exigencies of war opened the doors of the city hospital to all classes. In 1942, 242 maternity patients were admitted from other hospitals that could not accommodate them, including indigent, semi-indigent, and private patients. The public's acceptance of this move and the wartime swelling of the population spurred the Health Department to change its policy; in 1943 all services were opened to paying patients. As a result $22,099.19 was collected from patients in that year, compared with $13,490.59 in 1942. Yet, the hospital fumbled the ball by failing to institute proper machinery for collections. In 1947, for instance, only half the patients admitted to Gallinger were interviewed to determine their economic status. The advantage of the hospital's improved image and the opportunity to become more financially independent slipped further away when the hospital facilities were allowed to deteriorate so badly that private patients were seldom willing to go there. Patients who could pay were not always billed because of primitive accounting procedures. After new methods were installed in 1961, the hospital was able to double collections, and between 1963 and 1968 the hospital increased the proportion of total costs recovered from 6.9 percent to 16 percent.[47] Thus, the rise of third-party payers and the scarcity of beds in private hospitals forced city hospitals to admit some paying patients, but shoddy facilities tended to keep the numbers down and a penny-wise, pound-foolish attitude hindered collection. For many years the funds collected from private patients by city hospitals formed only a fraction of their income, even after the widespread adoption of hospitalization insurance, because of inertia, inefficiency, inadequate clerical staff, and the difficulties in eliminating the image of a pauper hospital.[48]

While city hospitals might welcome some paying patients, they were not as eager to use their facilities for another group of patients, the chronically ill and the permanently disabled. Private hospitals shut their doors on these patients, literally and figuratively dumping them on the city hospitals. They in turn shipped them as quickly as possible to isolated hospitals that cities had established for long-term care in out-of-the-way places—on nearby islands, as in Boston and New York, or in the far suburbs, as in Philadelphia, Chicago, and Washington. Doctors in training avoided hospitals for the chronically ill. Wards were usually crowded, poorly kept, and meagerly supplied; staff doctors were often medical hacks, ill-trained, uninspired, and uninspiring. Nurses likewise preferred hospitals for the acutely ill, where morale was higher and challenges seemed greater. As for the general public, once these patients were put away for long-term care, they were largely forgotten.[49]

Baltimore, though, managed the care of the chronically ill differently. Unlike hospitals such as Bellevue and Philadelphia General, Baltimore's city hospital retained sufficient land to provide for patients with both long- and short-term illnesses. Facilities for both the acutely and chronically ill, having remained together, suffered together while the hospital was the neglected stepchild of politicians; but when university services and better administration strengthened the general hospital, medical and nursing care on the chronic wards also improved somewhat. Yet, crowding continued; long waiting lists forced authorities to pack patients into every inch of space. During the Depression the board of supervisors tried unsuccessfully to persuade other Baltimore hospitals to share the burden of the chronically ill. In 1935 T.J.S. Waxter, director of the Department of Welfare, in order to learn about the hardships of these patients at first hand and to dramatize their plight, ate meals with the patients at the Baltimore City Hospitals and slept in the basement where some were housed. When the community was not responsive, the city hospital was forced to carve out of almshouse space the wards required to house patients needing long-term care. By 1940, despite these additions, the extended-care wards, listed as accommodating 480 patients, were 103 percent full, and by 1944 120 percent full. The predicament of many of the five hundred to six hundred patients awaiting admission was described as pitiful in the extreme. Some had to be placed in inappropriate institutions because "chronic patients must scramble to find refuge wherever they may." Moreover, even the best tended of the chronically ill patients at the city hospital were receiving only custodial care unless they became acutely ill.[50]

Into the breach stepped an unlikely champion. Mason F. Lord, scion of a prominent Baltimore family and a half-hearted medical student during his first years at Johns Hopkins, had caught fire while a medical resident at Baltimore City Hospitals. Placed in charge of the wards for the chronically ill in 1959, he fashioned a comprehensive plan for the care of these neglected patients. This included a medical and social evaluation for each person, leading to either rehabilitation or continuous care in an environment suited to their needs. Families were to be encouraged to care for their members if possible and to receive help in doing so, and other patients were to be sent to foster homes or nursing homes. Hospital beds thus freed were to be used for newly admitted patients so that they could be evaluated and properly placed. In his plan Lord was following methods evolved as early as 1925 at Montefiore, a private hospital in New York City. Its staff had shown that chronically ill patients could be divided into three groups: those requiring intensive diagnostic study, those requiring careful nursing, and those requiring only custodial care.[51]

Lord's zeal, his personal charm, and his social connections all helped to get the program under way. His enthusiastic staff, collaborating with other hospital departments and community agencies, placed many long-hospitalized patients with their families or in other suitable homes. For those who remained in the hospital, a program of rehabilitation and optimal activity was begun. Between 1963 and 1965 the average stay for long-term patients was reduced from 659.4 days to 205 days. Maryland had lagged in the development of satisfactory nursing homes (in 1960 only .48 acceptable beds per one thousand population were to be found in nursing homes in Maryland, compared to .99 for the United States as a whole).[52] Lord's program of placing patients where satisfactory care could be given served as a stimulus for the upgrading of these institutions in the Baltimore area. His untimely death in 1965 was a serious setback. Because he had barely gotten his program under way, his legacy was mainly conceptual and inspirational, but it affected the whole hospital and many medical and social agencies as well as the public of Baltimore.

Lord's program within the hospital itself did not fulfill its early promise, although his colleagues and successors did their best to carry it on. In the 1960s and 1970s the chronic hospital was modernized and the large wards converted to small and medium-sized rooms. Yet, a program of establishing units for special diseases such as arthritis, stroke, and orthopedic conditions foundered because salaries for nurses were not forthcoming. In fact, in 1965 fifty registered nurses were needed to care properly for the 505 patients on the chronic wards; instead there were only eight. At this time, however, city administrators had been ordered to hold the line on budgets, and the state inspectors looked the other way. Furthermore, it turned out that the hospital could not be reimbursed from Medicare funds for patients who reentered the chronic wards after developing exacerbations or complications of their illnesses. The resulting financial bind made city hospital authorities less interested in keeping patients under lifetime observation and treatment than had been contemplated.[53] Although the hospital did not succeed in meeting Lord's hopes in every respect, it had given a better life to hundreds of formerly neglected patients; it had shown that a city hospital could turn a depressing, dead-end existence for many of the chronically ill into years of healthier, happier living.[54] It is questionable whether these improvements would have been made so soon if long-term care had been provided in another, distant institution.

A few other city hospitals were also placing emphasis on the rehabilitation of disabled patients, and in each a particularly dedicated person or special circumstances provided the impetus. After the Second World War Howard A. Rusk of New York University built a nationally recognized rehabilitation institute and influenced the New York City Department of Hospitals to develop a rehabilitation program. Toward the end of the 1940s this department began to establish home-care programs for suitable patients and to transfer others to nursing homes and infirmaries (see Chapter 8). Rehabilitation programs were begun at the Rancho Los Alamos, the chronic hospital of Los Angeles County, after the poliomyelitis epidemics of the 1940s and 1950s. In the Wayne County General Hospital and

Infirmary of Detroit a citizens' group recommended in 1954 that each patient admitted to the infirmary be given a physical examination and investigated socially and financially, that a program of rehabilitation be started, and that the cooperation of social agencies and the state's medical schools be sought.[55]

A movement to give better care to institutionalized chronically ill patients was also developing. In New York City, for instance, following the leadership of Montefiore, long-term care in the city's institutions on Welfare Island was greatly improved. Boston's long-term facilities at Long Island Hospital were supplemented by converting its tuberculosis sanatorium at Mattapan into a hospital for the chronically ill.[56]

Another group of patients who sought the city hospitals in large numbers were members of racial and national minorities. In Boston, for example, as each wave of immigration swept more refugees into the city, some found their way to the city hospital. When members of a minority group became numerous enough to obtrude upon the life of other citizens outside the hospital, their presence in the hospital affected the attitudes of other patients, attendants, administrators, and staff and certainly of those who voted funds for the hospital. Prejudice or lack of interest often caused voters and lawmakers to oppose appropriations for hospitals that catered to new or unpopular minority groups. Civil servants were annoyed when their routines were disrupted by their inability to communicate with persons who spoke only what sounded like gibberish and followed alien customs. And ignorant attendants tended to view foreigners as did Jim in *Huckleberry Finn,* when he asks Huck, "Is a Frenchman a man?" "Yes," Huck answers. "*Well,* den! Dad blame it, why doan' he *talk* like a man?"[57]

In the city hospitals of the United States attitudes toward blacks illustrated the effect of number on prejudice in a community. In the hospitals of Boston and other northern cities where blacks were relatively few in number (see Table 6), they blended in with other groups; in most cities in southern and border states they were segregated either in separate wards or in separate hospitals. When segregation was practiced in a single hospital, deficiencies in the care of blacks were usually minor and subtle. In my experience at the city hospitals of Washington and Baltimore from the 1920s through the 1940s, similar care was usually given to members of all races by white house officers and nurses. Attitudes toward minorities were more likely to affect acceptance into the hospital than care within the hospital. When a minority patient breached the barriers and reached a hospital bed, the stranger became a patient, and the professional indoctrination traditionally given to doctors, nurses, and members of other health professions took over. Nothing obliterates personal distinctions more completely than a hospital gown. The professional saw mainly a problem to solve, a job to tackle, a goal to reach: achievement of the best possible health for that particular patient. Prejudice was usually subordinated to professionalism.

Staff attitudes up to the 1950s were characteristically kindly, helpful, and supportive. In fact, interns and residents preferred to treat black patients, because they usually cooperated more readily with medical and nursing staffs than many white patients. In this climate of benign paternalism most of the staff extended themselves as far and pushed themselves as hard to care for blacks as for the whites—but only so long as the black patient obeyed orders. Latent racial prejudices might take precedence over professional standards if a black patient were uncooperative. A Gallinger nurse later recalled that on such occasions "the attending doctors used to scold them and were pretty rough with them." And in some city hospitals in the South the lot of the black patient was harder. According to a staff physician in Houston's Ben Taub Hospital. "We are forced [by prevailing attitudes] to treat the poor Negroes as another species, test animals, relics of the Stone Age, not as sensitive as we to pain."[58]

The changeover in attitudes toward blacks in city hospitals mirrored changes in American society. The capital city and its public hospital especially reflected the nation's shifting sentiments. Early in the Civil War Congress emancipated slaves in the District of Columbia, forbade racial discrimination in railway cars, and opened public schools to blacks. In 1866 Congress voted unrestricted male suffrage in the District and overrode President Andrew Johnson's veto of the measure. When a territorial government was in operation in the District from 1871 to 1875, all citizens voted for members of the lower chamber of the local legislature, but when a commission form of government was substituted, elective officers disappeared. During the next quarter century the rights of blacks were steadily eroded in Washington and all over the nation. By the turn of the century Washington's city hospital was rigidly segregated, so much so that the superintendent reported in 1903 that he was forced to mix tuberculous patients with others because he did not have the rooms to isolate them and at the same time separate blacks from whites. Only white doctors, medical students, and nurses were trained in the institution. And nothing changed for nearly fifty years.[59]

Meanwhile, federally supported Howard University, which operated one of the two medical schools in the United States for training black doctors, had access to so few patients for teaching purposes that two interns were assigned to each patient in its only hospital, Freedmen's. In 1947 Federal Security Administrator Oscar R. Ewing, who had jurisdiction over the university, heard of these difficulties. He conferred with officials of Howard, Georgetown, and George Washington Universities and pressured the commissioners into granting Howard University a service at Gallinger Hospital where black house officers would be trained. Despite verbal fireworks from a southern congressman (which were ignored),* a controversy over transfusions of blood between persons of different races (which continued to be given),** and a proposal for separate living and dining areas for black doctors (which was rejected), the new service was inaugurated without incident in September 1948. Apparently patients, staff, and the public were better prepared for integration than some of their leaders. From the start it was decided that black doctors could treat white patients and vice versa. This was a step toward desegregation of the wards, which took place some time later. Yet integration was still incomplete; the nurses' training school did not admit black nurses until 1956.[60]

Integration of the races took place in Baltimore's city hospital around the same period, but with less fanfare. This hospital had maintained separate wards for blacks at least as far back as 1838, but separate did not necessarily mean equal facilities: in 1841 the hospital for black men occupied the upper floor of the former coal house. Doctors and nurses were drawn from the white population, although by 1927 one-third of the general hospital patients were black. In 1934, when many black as well as white nurses were unemployed, the National Urban League requested that the hospital hire some black nurses for black wards. The board of supervisors thanked them politely for their suggestion—and did nothing. But when it became difficult to find qualified nurses, the hospital employed black graduate nurses on wards for black tuberculosis patients and in 1950 on all wards. In that year also the Medical Advisory Board recommended that qualified black physicians be given the same opportunities as white doctors, and in the following year the first one was appointed. Finally, in 1956 a severe epidemic of poliomyelitis flooded the isolation wards, and patients could no longer be segregated according to race. Integration of these wards was accomplished so smoothly that desegregation of the entire hospital followed soon afterward.[61]

In the separate city hospitals for blacks the problem of integration

*Congressman John E. Rankin of Mississippi claimed that black doctors were being placed in Gallinger Hospital to humiliate white patients as a "cheap scheme of playing politics in order to get a few votes." U.S. Congress, House, *Congressional Record*. February 23, 1948, p. 1567.

**While negotiations with Howard University were going on, it was discovered that the hospital's chief of staff had rejected blood from the Red Cross because it contained blood from persons of all races. When his order was countermanded by the health officer, the chief of staff resigned. *Washington Post*. May 23, 1948, July 22, 1948; *Washington Star*. July 23, 1948.

was more complicated. Black leaders bowed to the need for their fellows to be hospitalized separately, maintaining that it was "infinitely better to be segregated in the hospital than out of it."[62] But they objected to being attended and nursed by a totally white staff and consequently welcomed separate hospitals as a training ground for black doctors and nurses. Black hospitals were run by two Missouri cities: City Hospital No. 2 and its successor, Homer G. Phillips Hospital, in St. Louis; and City Hospital No. 2 in Kansas City. The former had a white attending staff and a black associate staff, the latter a completely black staff. In the 1930s black leaders and several philanthropic foundations joined forces to persuade other cities to operate separate hospitals for blacks. In Cleveland, however, black leaders opposed a move to build a separate hospital as introducing segregation where it had not previously existed. But the proposal did serve to open wide the doors to black staff doctors and interns at the existing city hospital. Integration of this hospital was so successful that the president of the Cleveland branch of the National Association for the Advancement of Colored People reported in 1957 that it was the only hospital in the city "where integration as we understand the process has occurred."[63] In New York City, where blacks numbered about 343,000, or 5 percent of the population in 1930 and 1,144,000, or 14.7 percent of the population in 1960 (see Table 6), certain city hospitals, such as Harlem, became primarily black hospitals.[64]

While separation of races in the same hospital minimized the grosser forms of inequality, second-rate care was the rule in segregated hospitals. Black hospitals occupied the oldest buildings and used the most worn-out equipment. When the budget for the city hospital was tight—and when was it not?—the black hospital was called upon to skimp the most. Furthermore, by being separated in training, blacks were denied the opportunity to learn from superior teachers, both black and white. Finally, the black hospital was especially susceptible to being politicized. For instance, in the 1930s Kansas City Hospital No. 2 employed at least 95 percent of all professional and semiprofessional blacks in the municipal government. "It was thus misused and abused by the political faction currently in power" to appoint their supporters at all levels, including the superintendent and even some of the doctors.[65] The politicians' and the public's attitude toward black city hospitals was the same as toward all city hospitals, only more so: the patients had to be taken care of, but as cheaply as possible.

To improve patient care and training opportunities, black leaders began a drive to open both white and black city hospitals to staff and patients of both races. In 1955 both St. Louis City Hospital No. 1 and the Homer G. Phillips Hospital were integrated for staff and patients, and in 1957 the two city hospitals in Kansas City were consolidated and the staffs and employees combined for operation as a single, integrated hospital.[66]

Atlanta went one step beyond other southern cities in accommodating black patients. Since beds for blacks in private hospitals were pitifully few (as in so many cities), a separate building, the Hughes Spaulding Pavilion, was erected in 1952 as a part of the city hospital, Grady Memorial. The hospital operated this pavilion as a place where black physicians could treat their private patients and train black house officers, although neither black staff doctors nor house officers were appointed in the main hospital.[67]

Public pronouncements and paper appointments did not mean the end of second-class citizenship for black professionals. Integration of formerly black with white hospitals often meant loss of administrative, professional, and other positions by blacks, and some of the integration was "on perhaps a token basis." As late as 1963 black doctors and dentists in Atlanta were suing Grady Hospital for equal treatment in appointments at the hospital.[68] City hospitals were moving slowly and hesitantly toward equal rights for black professionals, patients, and employees; still their progress was no slower than in private hospitals. For instance, the board of commissioners, in abolishing racial discrimination in all units of the District of Columbia Health Department in 1953, led the way for larger voluntary hospitals to follow in 1954 to 1956.[69]

The event that greatly accelerated the movement toward integration in hospitals was a victory for civil rights in a court battle in North Carolina. The Hospital Survey and Construction Act (known generally as the Hill-Burton Act) had been passed by Congress in 1946, providing matching funds for construction of public and private, nonprofit health facilities. Although the federal agency administering the program had originally permitted funds to be used for separate facilities, a decision of the Supreme Court in March 1964 forced the Department of Health, Education and Welfare to require that discrimination could not be practiced in any part of an institution receiving Hill-Burton funds.[70] After this the barriers began to come down, although slowly in some places. Title VI of the Civil Rights Act of 1964 made all federal assistance programs subject to its nondiscrimination provisions, including funds paid through state and county health and welfare agencies for the care of indigent patients. But many private hospitals still barred black doctors from their staffs and thus effectively kept black patients out of their institutions. A study of fourteen communities around 1962 showed that whereas 71 percent of black doctors were affiliated with primarily white hospitals in Brooklyn, New York, only 28 percent were able to treat their patients in white hospitals in Philadelphia, 21 percent in Boston, and 7 percent in Chicago.[71] After the changes in federal regulations in 1964, many private hospitals continued this practice since they did not receive federal funds. This affected city hospitals profoundly, because blacks who were able to pay for private medical care through insurance plans were forced to use the city hospital instead.

The problem was especially acute at Cook County Hospital. How blacks were forced to use this hospital was demonstrated by the experience of a large Chicago trade union, 80 percent of whose members were black. Although their hospital expenses were covered by insurance, 23 percent of the members' illnesses were treated at Cook County Hospital in 1954, compared with 5.5 percent at Michael Reese, the private hospital with the largest number of black patients. A survey of patients admitted to Cook County Hospital in 1970 showed that many black patients were not poor; in fact, their median income was considerably higher than that of whites entering the hospital. Few were on public assistance, and twice as many blacks as whites had health insurance. But they were forced to use the public hospital since they were denied health services elsewhere because of their color.[72]

The Chicago Commission on Human Relations found in 1955 that 60 percent of all black babies born in Chicago hospitals were delivered in Cook County Hospital and that its share was increasing. Although one of every four children born in a Chicago hospital was black, only one of every twenty-five children born in a hospital controlled by a religious organization and only one of every nine children born in a hospital controlled by a nonprofit organization was black. Few predominantly white hospitals in Chicago granted staff privileges to black doctors. In 1960 it was reported that of 226 black physicians, only nineteen could admit their patients to primarily white voluntary hospitals. By 1965 the number had increased to sixty-four black doctors holding appointments in forty-two private hospitals other than hospitals primarily for blacks.[73]

Cook County Hospital was representative of city hospitals elsewhere; they were disproportionately filled with members of minority groups, especially blacks in southern and in northern industrial cities and Hispanic Americans in southwestern cities. These groups, of course, were the ones with the least political power, and the discrepancy between representation in the city hospital and in political office was to become an explosive issue in the next few years.

Alcoholics constituted another group of patients whom city hospitals were forced to take because they were not wanted elsewhere. Boston City Hospital had more than its share because of the prevalence of alcoholism among the Irish. Sears wrote of them in the

1920s: "On Saturday nights five patrol wagons at one time have been lined up before the door of the Accident Room, in which every available space has often been occupied by patients in all stages of alcoholic intoxication,—a noisy, filthy crowd of swearing, fighting, obscene individuals, who are kept partly in order by a detail of police." Most of those who were admitted needed only to sleep it off under observation, but a single noisy, uncooperative alcoholic could disrupt an entire ward. Accordingly, the staff sought separate facilities for them, and at times one of the older, unused wards was allocated for their reception.[74]

In 1930 the Fifth Medical Service was established and placed under the control of the Boston University School of Medicine. Within this service a basement ward was designated for alcoholics because the leaders of the service apparently saw this as a way to extend their unit, which was at first smaller than the other university services. In contrast to most facilities for alcoholics, this unit gave its patients the advantages of being treated by academically oriented doctors, led by Harold Jeghers. Their studies of vitamin deficiencies, which were common in alcoholics in the 1930s and 1940s, added much to the knowledge of these diseases and at the same time benefited the patients and stimulated the interest of doctors throughout the hospital in the complications of alcoholism. It was a prime example of making a virtue out of necessity, an approach that could have been imitated with profit in a variety of fields in all city hospitals.[75]

At midcentury the problems of the communities around them were increasingly impinging upon the city hospitals, but these institutions were so fully occupied in wrestling with their own problems that they had little time for outside ones. A few members of staffs and governing bodies had tried to relate their hospitals to the community. One of the most effective mechanisms for this purpose was hospital social service. Of all hospital patients, those who were in city hospitals because they lacked funds, friends, and influence needed contacts with the community and its support; yet they were scanted here as elsewhere. Budget makers, pressed to provide more doctors, nurses, attendants, secretaries, janitors, and repairmen, tended to put social workers at the bottom of the list. Although some city hospitals had pioneered in introducing social services (see Chapter 4), others remained sadly deficient. For instance, in the mid-1930s Gallinger had to depend on a few untrained volunteers, and as late as 1947 the hospital had only three social workers for 1,407 beds, while the Boston City Hospital had thirty-four for its 2,378 beds. At the Baltimore City Hospital a few social workers were paid from temporary funds at times for varying intervals; not until 1959 was a Department of Social Work permanently established.[76] Even in departments that had been operating for some time as well as in newer ones, the number of social workers was never enough to meet the need.

Others on the staff did what they could to help. Some doctors found time to deal with the social and emotional problems of their patients. They were impressed, as a Boston City Hospital staff member later wrote, by the "examples of human behavior at its very best and its very worst...Where life and death were only one bed apart,...[the city hospital] afforded...a vivid demonstration of what it meant to be a doctor." While Mason Lord was revolutionizing the care of long-term patients at the Baltimore City Hospitals, several doctors at that institution became so concerned with the psychosocial problems of the patients that they earned a reputation as the socially concerned part of the faculty of Johns Hopkins Medical School.[77] Some of their spirit lives on at the city hospital. In other city hospitals occasionally a doctor was inspired to minister to the personal needs of his patients, but in general the pressure of overwork amid shortages, confusion, and lack of interest in high places plus the technically oriented philosophy introduced by scientific medicine made attention to personal and social problems of patients seem like dipping a thimble into a sea of despair.

Much more than social service and a few concerned doctors was needed if city hospitals were to provide well-rounded medical care to all who required it. By the early twentieth century hospitals had become workshops for doctors and their helpers, where resources were assembled for diagnosing and treating patients. The machinery grew in amount and complexity, and the results in lives saved and health improved went beyond expectations. Yet, some medical services could be rendered as well or better without hospitalizing the patient. The most obvious of these was care of the ambulatory patient.

Early almshouses and the hospitals that arose from them were surrounded by a high wall, and at the entrance gate was a small house where conveyances bearing new inmates stopped. Although these appurtenances had disappeared in time, the attitude of separateness that engendered them remained. Also, poorhouses were usually set apart by being placed on the edge of the cities they served. Thus, the medical care of the walking poor was not the concern of the almshouse hospital. But as the residences of city dwellers multiplied in the vicinity of the city hospitals, the walking poor pressed on the hospitals' doors. In the large cities of England, beginning in the eighteenth century, dispensaries had been founded where volunteer doctors examined and prescribed for the sick who did not need hospitalization. Between 1786 and 1800 similar institutions were established in Philadelphia, New York, Boston, and Baltimore, and their number multiplied. By 1866 ten major dispensaries were operating in New York City. At first most outpatient clinics were independent, while some were attached to voluntary hospitals and a few were a part of large city hospitals. In the latter half of the nineteenth century, as hospitals became progressively more important in the health-care system, many opened outpatient clinics, while independent dispensaries either merged with hospitals or medical schools or disappeared. In 1926 a survey showed 1,252 outpatient clinics attached to general hospitals, fifty to medical schools, and 488 to specialty hospitals.[78]

Between 1944 and 1965 outpatient visits to general hospitals operated by local or state governments increased from approximately seven million to thirty million per year, or 310 percent (see Table 12). The rise in the number of visits to voluntary hospitals was nearly the same, 316 percent. After the Social Security Amendments of 1965 the increase in outpatient visits continued, but at a slower rate. Between 1965 and 1970 visits rose 20 percent in nonfederal government hospitals and 42 percent in voluntary hospitals.[79] Some patients continued to consult private physicians after the advent of Medicare, while others viewed hospitals as their family doctors. Most clinic patients patronized the voluntary hospitals, which were often closer to their homes and certainly more attractive.

Of all outpatient visits, those made to emergency departments of hospitals showed the most spectacular gain. They increased 205 percent between 1954 and 1965, 185 percent in local and state

Table 12. Outpatient visits to short—term, nonfederal hospitals in the United States, 1944–1979.

| Year | State and local Government | Percent increase | Voluntary | Percent increase | Total[a] | Percent increase |
|------|---------------------------|------------------|-----------|------------------|----------|------------------|
| All outpatient visits | | | | | | |
| 1944[b] | 7,308,964 | — | 14,248,733 | — | 23,666,033 | — |
| 1953 | 13,160,608 | 80.1 | 25,323,789 | 77.7 | 42,022,174 | 77.6 |
| 1965 | 29,961,641 | 127.7 | 59,233,208 | 133.9 | 92,631,497 | 120.4 |
| 1970 | 36,060,975 | 20.3 | 84,387,648 | 42.4 | 124,287,646 | 34.1 |
| 1979 | 54,059,742 | 49.9 | 140,524,663 | 66.5 | 203,873,394 | 64.0 |
| Emergency Visits | | | | | | |
| 1954 | 3,361,838 | — | 5,809,516 | — | 9,418,755 | — |
| 1965 | 9,588,234 | 185.2 | 18,309,938 | 215.1 | 28,733,158 | 205.0 |
| 1970 | 12,742,226 | 32.8 | 28,539,618 | 55.9 | 42,692,761 | 48.6 |
| 1979 | 19,396,699 | 52.2 | 52,454,667 | 83.8 | 76,634,443 | 79.5 |

Sources: Hospitals, Guide Issues, 1945, 1954, 1955, 1966, 1971; American Hospital Association, Hospital Statistics, 1980.
a. Includes proprietary hospitals.
b. Does not include a small number of visits to special short-term hospitals.

government hospitals and 215 percent in voluntary hospitals (see Table 12). After the introduction of Medicare, they continued to rise at a lesser rate: 49 percent between 1965 and 1970, again more rapidly in voluntary than in government hospitals (56 versus 33 percent). People had found that if they went to a hospital they could be seen by a doctor any time of the day or night, even though the wait might be long. A patient in the emergency room of the District of Columbia General Hospital, when asked why he had come, answered simply, "You know you're going to get a good doctor."[80] The hospital was becoming the general practitioner of the masses.

For a long time many city hospitals, overwhelmed by the burden of providing adequately for the patients on their wards, refused to admit, even to themselves, that services to outpatients were an important part of their obligation. Bellevue established an outpatient department in the 1860s, but another pioneer city hospital, Philadelphia General, did not open its first clinic (for genitourinary diseases) until 1919.[81] The city hospitals in Baltimore, Chicago, and Washington founded outpatient departments even later.

The rise of the outpatient department of Baltimore City Hospitals was typical of many almshouse hospitals. Situated peacefully amid its 140 acres and largely insulated from the rest of the city by industrial plants, the hospital was not disturbed by the needs of the walking poor, especially since provision for large numbers of outpatients had been made elsewhere. According to Alan R. Chesney, in 1930 223,653 visits were made to the Johns Hopkins and 98,675 to the University of Maryland outpatient departments and 101,767 to clinics maintained by the city welfare department. He listed Baltimore City Hospitals as treating no outpatients. In 1937 a critic noted that the new city hospital building contained no facilities for outpatients except for follow-up visits by patients recently discharged from the wards. Ten years later construction of a complete outpatient department was finally started; when it opened in 1950, it cared for only two hundred patients a day, although its capacity was five hundred.[82]

The sequence in Washington was similar. Ambulatory patients began to come to the city hospital after the modern general hospital building was opened in 1929. For many years they were treated in out-of-the-way places, wherever space could be found, including "an old frame structure graced with fallen plaster and, frequently, flooded floors." Not until 1947 was a complete outpatient department opened in a building specifically planned for this function.[83]

Cook County Hospital, though not evolving directly from an almshouse, nevertheless concentrated on the care of inpatients and formally ignored for a long time the needs of ambulatory patients. But convalescing patients had to be followed up after discharge, and others returned when a new illness developed; the hospital's doctors felt compelled to care for them wherever they could. The 1932 MacEachern report criticized the practice of holding outpatient clinics throughout the hospital wherever a spot could be found, because it was confusing and potentially dangerous to outpatients and ward patients alike. The county commissioners, in defending their methods, revealed that approximately two hundred thousand outpatients a year were being treated by this hole-in-the-wall method. With characteristic effrontery, they claimed that under their nonsystem the "present outpatient clinics cost the county only nursing service and medication" because the other services were performed by attending physicians and resident staff, "none of whom are on salary." They totally ignored the drain on inpatient wards caused by siphoning off doctors' services and the confusion, inefficiency, and danger produced by bringing patients from the outside into already overcrowded, undermanned, and germ-infested areas. Finally, when a citizens' committee renewed the criticisms, the commissioners acted. They purchased an abandoned private hospital nearby and converted it into an outpatient building. From 1940 until a more modern facility was opened in 1961, the staff cared for increasing numbers of patients there as best it could.[84]

The Boston City Hospital, unencumbered by almshouse traditions and emulating instead the Massachusetts General Hospital, contained an outpatient department from the start. In the year 1864, 371 patients were treated there, compared with 475 inpatients. In 1867 a separate outpatient building was opened and the staff enlarged, and in the following year nearly four times as many patients were cared for in the outpatient department as on the wards. The opening of the new building also provided space for clinics in medical and surgical specialties, which were clamoring for autonomy and Lebensraum (see Chapter 3). Outpatient clinics were established between 1868 and 1877 in dermatology, otology, gynecology, neurology, and laryngology (see Table 4). By 1903 60,730 patients were being treated in the outpatient department, compared with 11,490 on the wards; even more would have come if the staff had been willing to start an evening clinic. But doctors did recognize the growing importance of the outpatient department by recommending that it be open every day.[85]

In 1916 a committee of trustees and staff found much to criticize in Boston City Hospital's outpatient program: it was inadequate, with poorly arranged quarters, too small a medical staff, a "primitive" clinical record system, no arrangements for follow-ups, and no direct connection with the ward services. A new and larger building was opened in 1924, two more stories were added three years later, and other improvements were made. Still, this outpatient department did not achieve the status and prestige of the ward services. Rather, individual attending doctors were assigned groups of patients, as was originally done on the wards, and in the late 1940s house officers from various university services were required to merge into this system while serving their months on outpatient duty. In the 1960s Charles Davidson of the Harvard faculty initiated a system whereby patients discharged from a ward were assigned in the outpatient clinic to the house officer who had treated them on the ward. This system was also used by the other schools. The university services did not hesitate to dip into the outpatient clinic of the city hospital to study and treat patients with particular diseases. For instance, research groups in hematology and pulmonary and cardiovascular diseases connected with medical schools developed large and active specialty clinics when it served their purposes to do so.[86] But generally the medical schools showed less interest than they had for ward services because outpatient clinics in university and voluntary hospitals usually sufficed for their teaching. At the centennial of the Boston City Hospital, Dickinson W. Richards, himself the chief of a university service at Bellevue, scolded his colleagues for their neglect of the outpatient clinic:

> We are not community oriented in the broadest sense. We take our turn in the clinic but this means that doctors are changing, that care is discontinuous, that it is rare for a doctor to be able to have the time or the interest to provide a truly comprehensive study of a patient, and even rarer to sustain it. The out-patient...division ought to be the main center of communication between the hospital, the people in the community, the city services and the practicing physician; where the medical care of the community is coordinated; but it is not. It is more likely to be for the hospital staff a stepchild, for the patient an interim stop-over, and for the practitioner an unloading point.[87]

Whether medical schools were interested or not, the city hospitals were forced to respond to increased demands that began in the Depression, demands that caught them unprepared. Their usually crowded outpatient and emergency clinics were jammed; the usual waits were prolonged, sometimes by hours; the usual shortage of doctors, nurses, and attendants became so acute at times as to make a mockery of medical care.[88] Those hospitals that were already ill-equipped and ill-manned, which included most of the city hospitals, fell further below adequate standards of care. City hospitals generally tried to keep up with the added demands by hiring more full-time and part-time doctors in outpatient and emergency clinics, by filling their thinning ranks of house officers with foreign-trained physicians, and by building larger quarters for outpatient services.

Expanded facilities and added staff seldom met the increasing needs. At the District of Columbia General Hospital, for instance, the building that was opened in 1948 and pronounced a "splendid outpatient department" by a local newspaper in 1951 was filled with long lines of waiting patients a decade later.[89] In 1950 24,000 visits were made to outpatient clinics; in 1956 106,000. Patients waited for hours, and at the end of the day those who could not be treated were told to come some other day. The doctor in charge counseled them: "Direct your anger where it will do some good."[90] But city officials could have claimed the fault was not all theirs, since the city hospital was being called on to furnish 47 percent of outpatient services for the indigent population of the District of Columbia. In 1965 another larger outpatient clinic was begun and put in use the following year, and in 1968 an appointment system was installed to help eliminate some of the waiting, but crowding continued.[91]

The clinic building at Cook County Hospital that opened in 1961 had been designed to handle 250,000 visits per year. By 1965, estimating 325,000 visits for the year, the Chicago Board of Health described it as past the saturation point. More disturbing still, there were 438,000 visits each year to the admitting-examining room service. A few of these patients were admitted to the hospital, a few given appointments to the outpatient clinic (where the waiting period for nonurgent cases was said to be three months); but the majority were treated for the complaint that brought them and sent on their way with no provision for further care. In a well-run outpatient facility most of these patients would have been examined and treated initially and on subsequent visits in the outpatient department.[92]

Likewise, in 1965 at the Boston City Hospital the 280,000 visits to the outpatient department and 156,000 visits to the emergency clinic were straining these facilities beyond their capacity. When architects proposed that a larger building for outpatients be constructed at the end of a three-phase rebuilding program, community members on the building committee protested the delay, and an ambulatory services building was placed early on the agenda.[93]

By the mid-1960s the Baltimore City Hospitals had active ambulatory services with 120,000 visits per year to the outpatient department and sixty thousand to the emergency service.[94] Although stretched to capacity, this institution was spared the extreme crowding characteristic of many city hospitals by the presence of large outpatient clinics in the university hospitals, which were closer to the homes of most underprivileged people.[95]

The crowding of outpatient clinics and the injustice of requiring sick persons in lower socioeconomic classes to travel long distances to hospitals sparked a movement to bring medical services closer to the people served. Public health planners in the larger cities had dreamed for years of providing services to the underprivileged through neighborhood clinics. Child welfare clinics were established in many cities, beginning before the turn of the century and multiplying during the Depression years. By the 1960s there was considerable interest in citywide systems of neighborhood clinics for general medical care. Some of these clinics were established by city health departments and a few by universities (for instance, Columbia Point in South Boston, run by Tufts Medical School) or neighborhood groups (such as Yeatman Neighborhood Clinic in St. Louis).[96]

City hospitals, having as much as they could do to cope with the care of the patients already within their walls, were generally not interested in sponsoring such clinics. For instance, in Baltimore a comprehensive program for providing medical care to recipients of public assistance was established under the city health department. Plans included support to the patients' family physicians by hospitals acting as medical centers to evaluate each patient's initial physical status, to provide diagnostic and therapeutic services, and to hospitalize patients when necessary. Johns Hopkins and the University of Maryland Hospitals became medical centers when the program began in 1948, but the city hospital did not participate until 1953. In fact, so splintered was health care in Baltimore that the health commissioner in 1961 identified twelve programs of medical care, each administered separately.[97]

Chicago was a prime example of what happens when health care is left to individual institutions. As early as 1939 the *Journal of the American Medical Association* commented on "the absence of a broad, comprehensive, coordinated public hospitalization program adjusted to the needs of a total population of nearly four and a half millions." Cook County Hospital, although almost surrounded by areas of urban poverty and blight, practiced supermarket, come-and-get-it medicine. Community relations, such as they were, were left to the politicians. Even a minor outreach, the establishment of a branch hospital with inpatient and outpatient facilities in the southern part of the city, first proposed in 1949 so that the large underprivileged (and mainly black) population in that area would not have to travel for hours to obtain care at Cook County Hospital, was strongly opposed by Karl Meyer and the county commissioners. They claimed it would be cheaper to increase accommodations in the main hospital, but one suspects that they anticipated a loss of control if part of the hospital were placed elsewhere. Though Cook County Hospital was too unwieldy for good administration, its huge size made it ideal for building empires.[98]

By way of contrast, New York City had fourteen city-owned general and several specialized hospitals by the 1960s, which were located in various sections of the city.[99] In Washington, after the city hospital had been placed within the Health Department in 1937, the health officer planned to establish neighborhood clinics throughout the city. Some clinics were started, but they were never closely related to the city hospital (see also Chapter 8).[100]

In the mid-twentieth century cities were expanding, experimenting, and changing. But in general, city hospitals maintained the isolation imposed upon them earlier and in which they had become entrenched. Justifiably or not, policy makers in these hospitals felt that they should direct all their energies to caring for the patients who crowded through their doors. This slowness to change was a major reason why the protests that came after 1965 triggered explosions that shook these hospitals to their foundations, brought disaster to some, and led to radical changes in others.

In the period from 1910 to 1965 advances in the science of medicine and the education of doctors, nurses, and technicians produced remarkable improvements in medical care—in city as well as voluntary hospitals. Although administration was becoming professionalized, politics still intruded into hospital management. Lack of interest in city hospitals and unconcern over the fate of the sick poor kept budgets low, resulting in outmoded, inefficient, ill-kept buildings; insufficient personnel, supplies, and equipment; and low morale, leading one hospital official to exclaim, "The greatest achievement of this hospital in this fiscal year is that it managed to continue operation."[101] When a city's leaders displayed an interest, or when hospital leadership was exceptionally good, these deficiencies were lessened; much of the time, though, interest and leadership were minimal, and the city hospitals continued to crumble literally and figuratively, although most citizens did not notice until protests by hospital staffs exposed the defects for all to see.

# Part IV · *The Community Period*

# 8 · *Problems, Protests, and Possible Solutions*

> Despite the important and impressive role
> that tax-supported public hospitals have
> played and are playing in medical educa-
> tion, medical discovery, and medical care,
> they have come upon impossibly evil times.
> Ray E. Brown

> To survive, the public hospital must
> change. The public hospital as it is known
> today probably will not exist tomorrow.
> *American Medical News*

IN THE MID-TWENTIETH century the future of many large city hospitals looked bright. They had grown in size, in reputation, and in potential for skillful diagnosis and efficient treatment. They were supported externally by affiliations with medical schools and internally by highly professional nursing schools. Despite persistently low budgets, leading to shortages in supplies, equipment, and nurses and other trained personnel, morale was usually high, internships and staff positions were generally in demand, and many competent administrators, doctors, nurses, and others willingly dedicated a lifetime in their service.

Yet, now, less than two decades later, these same devoted workers might echo King Lear—"We have seen the best of our time"—for the picture had changed drastically. Protests, charges and countercharges, marches, strikes, and sit-ins by doctors, nurses, and concerned citizens had at times reduced the effectiveness and had even threatened to close many of these institutions.[1] What had brought city hospitals to this pass?

In the 1960s many medical students in the United States joined in the outcry against injustices to minority groups and the poor.[2] Later, when some of these students became house officers in city hospitals and had to cope with the deficiencies of these institutions, they abandoned the customary eyes-front behavior of house officers and joined other groups in rebelling against the quality of health care being provided the sick poor. Prevailing attitudes were summarized in a 1973 conference on hospitals: "The poor who get care do not like the public hospitals; the boards of supervisors or city councils who have to raise taxes to support them do not like them; the people in general do not like them because they are stigmatized as providers of second class medicine; taxpayers resent their taxes going into facilities for the poor; and health planners want them to go away."[3]

Those who ran city hospitals first denied the charges against their stewardship and then tried palliative remedies; when these failed, they began looking for root causes. Cook County Hospital's recent history has been an example of such a revolt and the hospital's attempts to respond. In the past few decades an intern had survived in this hospital by becoming a lone wolf, attending to his own patients, furthering his own career, and ignoring everything else, and nurses had lived lives of quiet desperation. The disaster syndrome had prevailed (see Chapter 7). But in the 1960s the full-time medical staff, which the authorities had built up as an alternative to affiliation with medical schools, appropriated one of the prerogatives of academicians: the right to speak their minds. They became increasingly vocal in demanding sufficient numbers of nurses, attendants, and laboratory workers and enough supplies and equipment for adequate medical care. Sometimes their protests succeeded in getting appropriations augmented, but the increase was never enough to remedy the principal defects.[4]

In 1969 Governor Richard B. Ogilvie, formerly president of the Board of Cook County Commissioners, informed by the staff and others about the hospital's problems, persuaded the state legislature to remove the institution from the direct control of the commissioners. Prodded by a full-page newspaper advertisement purchased by the hospital's interns and residents, which announced, "The Hospital is Dying and the Doctors Are Helpless," the legislature established the Health and Hospitals Governing Commission, composed of nine citizens appointed by the governor to administer Cook County's hospitals and health department. When the legislation proved too vague to give the commission sufficient powers, a coalition of groups representing blacks put pressure on the county commissioners, while seventy staff doctors and most of the hospital's house staff and nurses threatened to resign. At the last minute the legislature passed a bill strengthening the powers of the new commission, which took control of the hospital on July 1, 1970.[5]

James G. Haughton was brought from New York City's Health Services Administration, where he had been deputy administrator, to be executive director of the new commission. He moved swiftly to reduce the size of the hospital and upgrade some facilities, and the hospital succeeded in obtaining full accreditation once again. But Haughton, who was accused by some of arrogance and highhandedness, was soon locked in a struggle with the hospital's full-time staff over professional standards and the collection of fees. Many house officers resigned, and the quarreling and bickering so damaged the hospital's reputation that house staff positions could not be filled with American-trained doctors. The battle reached a climax in late 1971, when the governing commission fired five staff doctors for allegedly threatening to "close and destroy" the hospital. By the time the doctors had been cleared of this charge by a court-appointed panel, morale had plummeted to a new low. It was not improved when an eighteen-day strike by house officers seemingly accomplished little except to bring a fine of $10,000 upon the house officers' association and ten days in jail for their leaders. Nor had the commission learned its lesson, for it compounded confusion by discharging the chief of medicine for allegedly aiding and abetting the strikers. Although he was reinstated by the court, the squabbles at the hospital, which were thoroughly aired in the news media, disgusted a large segment of the Chicago community.[6]

Having beaten their plowshares into swords, administrators, county commissioners, and doctors continued to fight. By 1978 the cacophony had reached a crescendo. The hospital had to borrow to meet its payroll, the governing commission filed a suit against the county board, and doctors and nurses agitating for increased appropriations were arrested during a sit-in in the board's office. The decade of discord ended in December 1979, when the legislature abolished the Health and Hospitals Governing Commission and returned the hospital to the jurisdiction of the county commissioners. The experiment in citizen control had failed.[7]

Depending upon one's point of view, this debacle might be blamed on personality conflicts, overrigid, insensitive administrators, obstreperous doctors and nurses, an indifferent medical profession, or a government with insufficient funds to pay its bills, and no doubt each of these contributed. Yet, much of the blame must be placed upon the people of Cook County and their leaders, in and out of government, who sanctioned a century of maladministration in which self-interest took precedence over the common good—the care of the sick poor. Chicago had sown the wind and reaped the whirlwind.

Public protests and financial stringencies in the 1960s and 1970s also forced other cities to scale down the size of their publicly owned hospitals. Such a task could never be easy or pleasant, but at least one institution, the Boston City Hospital, accomplished it in a more orderly manner, without explosions and with a less bitter aftertaste. In 1964, when the hospital celebrated its centennial year, the *New England Journal of Medicine* contrasted its three original departments with the current twenty-one clinical and three laboratory services; its original thirteen attending doctors with the current staff of four hundred physicians and surgeons; and its original five interns with the current house staff of 349. Two medical schools, in addition to the original affiliate, Harvard, had become associated with the hospital during the intervening century, and the editor praised the equal division of services among these schools as a healthy balance of power.[8] Yet, a visitor looking beyond the dedication of the doctors and nurses could see that the hospital was crowded, shabby, and run-down.

In the past the staff had tried to cope with the institution's defects by protesting to authorities but without exposing the deficiencies to public view. Finally, concluding that extreme needs require extreme remedies, they broke their silence. In 1968 they announced that the hospital was "reaching the end of the line as a functioning medical unit." They recounted shortages and cited telling examples of the results: premature babies dying for lack of nurses to feed them at required intervals; a night when fifteen wards of twenty or more patients were without a single nurse; and the unavailability of laboratory tests in emergencies for lack of equipment or technicians. When these public warnings went unheeded, the ultimate indignity followed: loss of accreditation in 1970. Although the institution was accredited on a provisional basis within five months, it was obvious that this once model hospital had now joined other city hospitals, running on a treadmill to stay in place.[9]

City authorities had also realized that the hospital's future was precarious. Even as the centennial was being celebrated, a committee had been studying the hospital's role at the mayor's request. The committee recommended creation of a Department of Health and Hospitals to coordinate the work of the Boston City Hospital, the two chronic hospitals, and a network of community clinics to be established throughout the city. The new department would be run by a commissioner and a board of citizen trustees appointed by the mayor. The new plan was initiated in 1966, and by 1968 an up-to-date laboratory for the diagnosis of infectious diseases, the hospital's first new building in thirty years, had been opened, and planning had started for a huge renovation and construction program. Integration with the chronic care hospitals was begun, and partial success was reported in persuading private hospitals to participate in the planned citywide network of clinics and hospitals.[10]

Meanwhile, although the hospital's administrators were making short-term improvements to regain full accreditation, the house officers' association protested that these were inadequate. They charged that most of the nine trustees were political appointees, including four who held jobs in City Hall, and that the city was running the hospital as much as possible on money earned by the hospital and appropriating only minimal amounts from tax funds even though grave deficiencies still existed. In 1972, a few days after the worst defects had been remedied sufficiently to gain the hospital full accreditation, the mayor announced that the city would reduce its appropriation to the hospital by almost 20 percent and that he planned to make the hospital support itself within five years. After much agonizing, the hospital's trustees met this budget slash by decreasing the number of house officers by 20 percent, eliminating the Harvard and Tufts services at the hospital (see below), closing the nursing school, and projecting a gradual reduction in capacity from 850 to five hundred beds.[11] The city authorities, like a doctor weaning a patient off an addictive drug, were progressively decreasing the hospital's supply of funds. Their prescriptions may have been rational, but the question in the minds of the hospital's own doctors was how much reduction the patient could stand.

Furthermore, although combining public health services and agencies for rendering medical care was supposed to achieve better coordination, it also added more layers of bureaucracy and could increase rigidity, add red tape, and impede action. Many observers within the Boston City Hospital believed that this is what happened when the Department of Health and Hospitals was created. Similarly, in Washington the bureaucracy of the Department of Human Resources became so unmanageable that the forty-year association of health department and hospital was dissolved in 1977, and the hospital was placed under a separate board of eleven commissioners nominated by the mayor.[12]

These examples illustrate the complexity of the problems of city hospitals in the 1960s, the intensity of the protests, and the difficulty of achieving solutions. Measures proposed for extricating these hospitals from their difficulties focused on the causes of the problems, which fell into three main categories: an inefficient, bureaucratic, politically handicapped administration; alienation from the community; and insufficient funds. The solutions that were advocated varied widely. They included remedying the defects of city hospitals without changing their policies or organizations; placing some or all of the administration in the hands of private, nonprofit organizations that existed to serve the public interest; taking positive steps to relate city hospitals more closely to the communities they served; and abolishing city hospitals altogether.

Some believed that satisfactory results could be achieved if city hospitals did what they had done in the past but did it better. This seemed reasonable, because reductions in size should make the task easier. Between 1950 and 1970 the combined capacities of the city hospitals of Baltimore, Boston, Chicago, and Washington decreased by 44 percent and in the next seven to nine years by an additional 26 percent (see Table 7). But cumbersome administrative machinery, outmoded and deteriorating physical plants, and workers with civil service or political holds on their jobs remained behind to add to costs and inefficiency. Administrators at Boston City and Cook County Hospitals discovered this when they attempted to streamline operations as they decreased capacity.

One ingenious method of mandating efficiency was to place a hospital under court surveillance. In the late 1970s house officers at District of Columbia General Hospital, following the example of their colleagues at Los Angeles County Hospital, persuaded patients to institute a class action suit to force improvements in medical care at the hospital. As a result, the court undertook to monitor the administration of the hospital for five years. But such an expedient, while it might motivate administrators temporarily to do a better job and generate public pressure for an increased budget, could not get at root causes and would actually add another layer of management.[13]

Nothing succeeds like success, and from midcentury on the medical schools appeared to thrive and were generally applauded. Since the portions of city hospitals controlled by medical schools tended to be managed better than some other areas, it seemed reasonable to call on the schools for help. In some hospitals individual services, such as laboratory examinations, were contracted out to universities. In others, a medical school was awarded a contract to provide all of the medical care in a city hospital.[14]

In establishing closer ties to medical schools, hospital authorities became convinced that affiliation with a single school was optimal administratively. When Trussell developed his affiliation plan for municipal hospitals in New York City, beginning in 1961, he followed this pattern, and when the decreasing patient population of Bellevue Hospital made a single affiliation desirable, Cornell and Columbia withdrew, leaving New York University as the sole affiliate by 1967. This enabled university and hospital to carry forward an integrated program of construction and rebuilding and to develop a contractual arrangement whereby New York University provided the services of physicians and other personnel and supervised the medical staff and training of house officers in return for semiannual payments, which in 1971–1972 amounted to nearly $8 million. Because Columbia agreed to affiliate with nearby Harlem Hospital instead of Bellevue, and Cornell preferred to be affiliated with voluntary general hospitals, the loosening of these historic ties was accomplished without hard feelings and attracted little attention nationally.[15]

In contrast, when the Boston City Hospital in its reduced size could accommodate only one medical school, the choice was difficult and stressful and had to be made in the public eye.

By the late 1960s each of the three medical schools affiliated with the Boston City Hospital had prestigious teaching programs for students and house officers and nationally known research programs. In 1968 the hospital paid out $821,663 in salaries for physicians, compared with nearly twice that amount paid by the universities: $442,638 by Boston University, $681,861 by Harvard, and $415,233 by Tufts. These doctors spent part of their time on hospital business— patient care, teaching of house officers, and administration—and part in teaching medical students and performing research. The last two activities increased space requirements and the hospital's overhead. In addition, city officials had wondered for some time whether tripling several diagnostic and treatment facilities might not be costing the hospital more than the services of three medical schools were worth, although some representatives of the schools contended that duplication of facilities was infrequent and when it did occur was more than offset by the added resources furnished by the schools. When expenses continued to mount while the number of patients treated was decreasing, and the presence of three leading medical schools had not been sufficient to prevent the ignominy of loss of accreditation, the authorities decided to limit affiliation to one school. To many, Harvard seemed the natural choice because of its 109 years of service to the hospital, the high repute of its services, the quality of its staff's research, and the amount of money it provided. Instead, when the authorities made their decision in 1973, after much deliberation they selected Boston University, reasoning that its proximity to the hospital would facilitate integration of specialized services and that it could best join the city hospital in caring for the large poverty areas surrounding the hospital. Thus, since all three schools had well-balanced services for medical care, teaching, and research, the hospital authorities chose primarily between a larger, more prestigious in-hospital program such as Harvard offered and a wider-than-hospital program that many people believed represented the wave of the future.[16]

In retrospect, one wonders whether Harvard would still be affiliated with the Boston City Hospital if it had heeded the advice of its dean, George P. Berry, given at the hospital's centennial seven years before. He stressed the need for community services affiliated with a medical school as extensions of the city hospital and warned: "University Medicine must adjust—and adjust promptly—to changing patterns of medical need and to the strident demands being heard on all sides for more and better care...Indeed, university medicine must exercise leadership."[17]

Sometimes the decision to leave a city hospital was made by a university. For instance, George Washington University Medical School decided in 1974 to end its forty-five-year affiliation with District of Columbia General Hospital because it preferred to teach in other hospitals rather than continually confront the multitudinous problems of the city hospital.[18] And it will be recalled that the University of Maryland withdrew from an even longer association with Baltimore City Hospitals in the 1960s. Both of these decisions were easier to make because voluntary hospitals had changed in the preceding century. These institutions, once only tolerating the presence of medical students, had come to welcome students, faculty members, and research teams. Correspondingly, the attitudes of private patients had changed so that they understood the value of being subjects for teaching and clinical research.

The use of private patients for teaching medical students increased gradually as the costs of hospitalization were paid more and more by insurance and was greatly accelerated after 1965, when hospitals were compensated through Medicare and Medicaid. Adding to the shift of clinical teaching away from city hospitals was the upgrading of the hospital services of the Veterans' Administration after the Second World War. As older veterans' hospitals were modernized and new ones built, and as ties with medical schools were tightened, these hospitals often became more attractive than city hospitals to the schools. By 1970, in the cities of Baltimore, Boston, Chicago, and Washington, three-quarters of the patients used by the medical schools for teaching were in private, Veterans' Administration, or university hospitals, and only one-quarter were in city hospitals (see Table 13). This ratio contrasted with approximately half in each category in 1920. In the intervening half century the number of patients available for teaching in city hospitals affiliated with the medical schools of these four cities decreased by 13 percent, and the number available in affiliated private, Veterans' Administration, and university hospitals increased by 144 percent.

Table 13. Beds used for teaching of medical students in Baltimore, Boston, Chicago, and Washington, according to type of hospital.

| | 1920 | | 1970 | | Change | |
|---|---|---|---|---|---|---|
| Type of Hospital | No. of beds | Percent of total | No. of beds | Percent of total | No. of beds | Percent change |
| City- or county-owned | 6,412 | 48.6 | 5,566 | 25.2 | −846 | −13.2 |
| Private, Veterans' Administration, and university-owned | 6,782 | 51.4 | 16,550 | 74.8 | 9,768 | 144.0 |
| Total | 13,194 | 100.0 | 22,116 | 100.0 | 10,978 | 67.6 |

Sources: American Medical Association, Council on Medical Education, "Hospital Service in the United States," Journal of the American Medical Association 76 (April 16, 1921):1089, 1090, 1092–1093; Hospitals, Guide Issue 45 (August 1, 1971):49–50, 64–73, 97–106.

Where medical schools remained affiliated with city hospitals, the partners tended to come closer. Some believed the schools should be given more administrative control. Varying amounts of responsibility for running city hospitals were transferred to medical schools in several cities such as Cincinnati, Rochester (see Chapter 6), Seattle, and Los Angeles, and in general these programs seemed to work well. In 1961 New York City began its gigantic experiment of affiliating each municipal hospital with either a medical school or one or more voluntary hospitals. The results are hard to evaluate because of extraneous factors—the advent of Medicare, changes in city administration, and the city's financial crises—but it appears that improvements in the hospitals occurred in those areas where the medical schools had interest and expertise. Progress was slower in outpatient departments, in programs for care of the chronically ill, and in home care, reflecting in part the lesser interest of academic institutions in these areas and in part the general lack of innovative methods for solving problems of providing health care beyond the wards of general hospitals.[19]

A more extreme expedient for avoiding the complexities and rigidities of administration by city governments was to form a public corporation to administer city hospitals. In 1970 New York City replaced its Department of Hospitals with a public benefit corporation, the Health and Hospitals Corporation. Its board of directors was composed of five members appointed by the city council, five by the mayor, and five city officials. In 1970 this corporation took control of a $617 million budget for eighteen municipal hospitals containing sixteen thousand beds. As this voyage to a new world began, the city was heading into a period of financial troubles, and the course of the ship has been stormy thus far, it having changed skippers several times.[20] Chicago's venture on the high seas, the Health and Hospitals Governing Corporation, found the going even rougher, and the new ship having made little headway in ten years, the crew was ignominiously discharged. Perhaps before expecting too much of a change in government, those who are concerned with the future of city hospitals might consider the breadth and depth of the problems and search for more far-reaching remedies, while heeding Pope's dictum:

For forms of government let fools contest
Whate'er is best administered is best.[21]

One glaring deficiency of city hospitals was their reluctance to reach out into the community. Perhaps more than anything else this contributed to their alienation from the communities they served. Certainly the people were making their needs felt. For instance, they came to outpatient departments and emergency clinics in increasing numbers. In the fifteen years following the advent of Medicare in 1965, the demand upon city hospitals for these services nearly doubled (see Table 12). Yet, these clinics continued to be crowded, undermanned, underequipped. They stressed episodic rather than continuous care—patch-up treatments rather than preventive medicine. They integrated poorly with the activities of families, neighborhood and racial groups, and community institutions.

Community relations were handicapped by the lack of political power on the part of the underprivileged classes of the cities. In Boston, for instance, political leaders tended to represent the groups that in large part had moved to the suburbs and were not very responsive to the Spanish-speaking and black constituencies. Furthermore, federal programs that provided for the establishment of clinics to serve specific areas of a city, beginning in the 1960s, tended to channel patients, funds, and public interest away from city hospitals. For instance, in Chicago the Mile Square Health Center, developed for the purpose of supplying health services to residents of one area of the city and funded by the Office of Economic Opportunity, was initiated and administered by Presbyterian–St. Luke's, a voluntary hospital. In Boston, however, the city hospital had committed itself by 1970 to develop programs of primary medical care in five areas of Boston with underprivileged populations.[22]

Although hospital administrators and staffs had little experience in relating to communities, some hospitals tried various expedients. One procedure was to appoint community representatives to governing boards. When this was tried at the Boston City Hospital, some questioned its efficacy, believing that these representatives tended to overemphasize their roles as advocates for their communities rather than identifying with the hospital. A more successful venture, in the same institution, was the formation of seven advisory committees with members from board of trustees, staff, administration, and community. At first community members were handicapped by their unfamiliarity with the hospital's operations, but as they gained experience they made their needs understood, as when they persuaded the building committee to move the projected outpatient building to the top of the list for new construction (see Chapter 7).[23]

More fundamentally, a city hospital could combat community alienation if it helped the community more. For instance, New York's city hospitals developed a program of home-care services as early as the 1940s. By 1958 these hospitals were providing 774,107 days of home care, 95 percent of the home care provided by all the hospitals in the city. Unfortunately, these services declined somewhat in the next few years as affiliations with medical schools were augmented and attention shifted to strengthening inpatient services with emphasis on full-time attending physicians, more competent house officers, and building research teams. Extending medical care from the hospital into the community was one reason for joining city hospitals with health departments in a comprehensive city department. But in many cities hospitals and community programs continued to go their separate ways. As late as 1979, for example, Washington had fifteen neighborhood health centers, none of them linked to any hospital.[24]

Denver was an exception. Beginning in 1966 the Department of Health and Hospitals began to build a network of neighborhood health stations for primary health care, backed up by several area health centers that provided specialized medical, laboratory, and other professional services. All were coordinated with the new Denver General Hospital and the University of Colorado Medical School. Aides for the hospital and the health centers were recruited from adjacent neighborhoods.[25]

City hospitals were loath to commit the full range of their services to a community health program for fear of being swamped, though it could be an effective way of assuring health care of high quality to an underprivileged population. A promising move in this direction was made in 1970 when the Los Angeles County–University of Southern California Medical Center established the Med-8 complex. Four contiguous wards were set aside in the hospital for the reception of people from a poverty area in the city. The same professional personnel served these patients when they were ambulatory or hospitalized, and home care was provided under the management of the same unit.[26] The results of this experiment are being watched with great interest.

Baltimore's city hospital served its surrounding community differently. It was in a unique position because its neighbors were primarily blue-collar workers in nearby industrial plants, while the city's indigents lived nearer other hospitals, especially Johns Hopkins and the University of Maryland, which provided them with extensive inpatient and outpatient services. In 1972, because the patient population of the city hospital had been declining, physicians and dentists at the city hospital formed a nonprofit corporation, Chesapeake Physicians Professional Association, which grew from fifty-eight full-time doctors to become in six years Maryland's largest group practice of medicine, with 110 full-time and approximately one hundred part-time doctors. The group renders medical and dental care to all patients in the city hospital and its outpatient clinics and the inmates of the city jail under contracts with the city government and to three health maintenance organizations, which give medical care on a prepaid basis. Since all doctors are members of the Johns Hopkins faculty, recruiting has been done jointly by the corporation and the medical school; thus, the group has attracted good doctors and a good house staff. The group has bolstered the finances of the hospital by filling beds and increasing income, and it has even added new services without raising the amounts of the annual contract.[27]

In 1970 a survey of patients' attitudes toward the Baltimore City Hospitals revealed that most of them rated medical, nursing, and overall care very high. One-fifth believed the hospital was better than other hospitals in the city, and over half thought it was as good. Four-fifths said they would continue to patronize the hospital. Apparently, preventive procedures had not been neglected since three-fifths of these patients reported receiving complete health checkups during the preceding two years, many of them having been performed in the outpatient department. A sampling of nearby residents who had not been frequent users of the hospital showed that the great majority expected to use it when the need arose. Physicians in private practice in the neighborhood rated care at the hospital high and believed that the institution should become more involved in providing health care to people in the area. Finally, staff executives of eighteen public and fourteen private agencies in the Baltimore metropolitan area had strongly favorable attitudes toward the hospital. Nearly all reported that the needs of the community and of their agencies' clients were being met adequately by the hospital. The surveyors concluded that "the record of this institution is very impressive."[28]

Another way for a city hospital to relate to its community is to arrange for community doctors to care for their patients in the hospital. One of the principal objections to placing the professional services of New York's city hospitals under medical schools was the elimination of neighborhood practitioners from the staffs and the consequent loss of an important link with the community. Baltimore's city hospital tried offering staff positions to neighborhood doctors to enable them to care for their private patients, but the program failed because staff membership necessitated appointment to a medical school faculty and teaching of medical students, which only an occasional neighborhood practitioner was willing to undertake.[29]

These three programs represent ways in which a few city hospitals tried to relate to their communities. In contrast to some other city hospitals, they were heeding the warning of the Commission on Public-General Hospitals that their future depended on "their ability to become broad-based community resources, providing essential services that contribute to a continuum of care within rationally planned and organized health care delivery systems."[30]

Some observers contend, however, that resolving the day-to-day problems of a city hospital, relating it more to the community, and arranging for an outside agency to administer part or all of the hospital are not enough. They believe that local governments do not need to own general hospitals and that the present city hospitals should become independent community hospitals in order to serve communities better, eliminate cumbersome bureaucracies, and free city governments from the responsibility of running hospitals. Baltimore seems to be moving in this direction. In November 1980 the citizens passed a referendum giving the city council authority to convert the Baltimore City Hospitals into a community hospital if desirable.[31]

. The idea of getting cities out of the hospital business had begun to be widely discussed in the 1960s. Formerly many a battle-weary city official or hospital administrator may have longed for such a solution to his problems, but he had usually kept his wishes in the realm of fantasy. Now it appeared to be possible. The growth of health insurance and the advent of Medicare and Medicaid dangled before city hospitals the possibility of paying their own way. At the same time the flight of the upper and middle classes to the suburbs was diminishing the tax base of the cities and of whole metropolitan areas and, by widening the distance between the miseries of the inner city and the comforts of suburbia, was making it harder to extract money from the comfortable to care for the miserable. What better solution than to give the costly, troublesome, constantly complaining, unattractive, unloved brat to somebody else to rear? Even the politicians, never likely to give anything away without something in return, were attracted to the idea.[32]

Between 1964 and 1974 five major cities—San Diego, Seattle, Kansas City, Toledo, and Newark—transferred their hospitals to state or private authorities in close association with a state-owned medical school. The municipality in each case contracted for services for its medically indigent citizens. The motivation for these changes seems to have been partly a desire to improve administrative flexibility and efficiency but mainly a desire to shift more of the increasing and unpredictable costs to the state.[33]

In New York City some municipal hospitals were closed altogether. In 1968 the City Club had published a pamphlet entitled "An End to Charity Medicine," in which it had recommended "a single system in which all hospitals would become voluntary (non-profit) hospitals." Soon after the Health and Hospitals Corporation began operation, mounting costs forced its administrators to project a deficit for 1972 of approximately $100 million. To save money, the corporation's officials proposed closing two hospitals, Sydenham and Gouverneur. Although the move was strongly opposed by medical staffs, hospital workers, and neighborhood groups, the officials persisted, and by 1977 four hospitals and parts of two others had been closed.[34]

New York's city-wide system of public hospitals enabled other municipal hospitals to absorb patients from those that had been closed down. Philadelphia, on the contrary, had only one city-owned general hospital. Around 1914 a proposal to split the Philadelphia General Hospital into several regional hospitals had been opposed by the staff and allowed to die. Instead, a huge institution piled up at the Blockley site. Thus, a half century later, when the run-down buildings required expensive renovations if the hospital were to continue operation, while at the same time the city's revenues were shrinking, the city owned no other suitable institution in which it could care for patients. In February 1976 Mayor Frank Rizzo directed that this historic hospital be phased out over the next eighteen months. Protests by hospital workers, staff, and community groups followed, but they were not as severe or as prolonged as many had expected, especially after the courts confirmed the legality of the city's actions. The Philadelphia General was formally closed as a hospital for acute illnesses in June 1977. To take over its functions the city established family medical centers in various sections of the city for ambulatory care of underpriviliged patients and contracted with five hospitals, including those associated with the University of Pennsylvania and Hahnemann Medical College, to staff the centers with physicians and provide hospitalization and back-up specialty services. Provision was also made for services usually furnished by city hospitals. For instance, treatment for victims of sexual assault was supplied by two university hospitals. Arrangements were also made for the care of the chronically ill. Admission of nonpaying patients to private hospitals presented a larger problem, since the city government had made no provisions for reimbursement. It seemed by 1981 that the new system had not yet been subjected to an extensive test, such as an excess of illness in the city or a large epidemic, and some observers wondered whether the many patients who formerly would have gone to the city hospital were failing to seek medical care. Nevertheless, most agreed that the transition had worked more smoothly than had been expected.[35] A major reason may have been that Philadelphia, like Baltimore, was relatively well provided with medical facilities, including five medical schools, each with a university hospital.

Whether or not to close public-general hospitals is a dilemma facing a number of large cities in the eastern and midwestern United States, where outmigration is the rule and decay in the central cities seems to be accelerating. In contrast, in the West and Southwest populations are increasing and cities are continuing to build. City hospitals in Los Angeles, Denver, and Seattle, for instance, seem to be able to keep up with needed construction and to supply reasonable amounts of equipment, supplies, and auxiliary help so as to provide good and sometimes excellent medical care. Their situations are reminiscent of

Boston City, Bellevue, Baltimore City, and District of Columbia General Hospitals twenty or thirty years ago.

While abolition of publicly owned hospitals might work in a few cities, there are good reasons why the larger cities at least should continue to maintain a public-general hospital. The past record of voluntary hospitals in caring for the indigent has been spotty. In good times these institutions have held out a hand to the poor; during hard times, when their help was needed the most, they have tended to close their doors to those who could not pay. In times like the present, when payments by patients, insurance carriers, and government are meeting costs, and especially when beds are waiting to be filled, many of them accept patients from lower socioeconomic strata. Today the private hospitals as a group might favor eliminating locally owned public-general hospitals, but when the gap between costs and third-party payments widens, or when funds from federal and state governments decrease, they are likely to wish the public hospitals had remained. The voluntary hospital system grew up along with the public hospital system through a de facto arrangement whereby the latter absorbed the patients the former rejected. Whether the private system can care for all patients is questionable. In 1969, for instance, private hospitals in the District of Columbia acknowledged their dependence on a city-owned hospital when six of them pledged to lobby in Congress for emergency funds to enable the District of Columbia General Hospital to survive.[36]

Voluntary hospitals are ill-prepared to care for particular groups of patients, either because their facilities cannot be readily adapted for the purpose or because of social distinctions that are not likely to be eliminated. These groups include, first of all, patients with problems that involve the public health or safety, such as patients with contagious and venereal diseases, alcoholics, drug addicts, and victims of trauma, burns, and sexual assault. It is doubtful that appreciable numbers of these patients would be welcome in most private hospitals. Second, when large-scale catastrophes or epidemics occur, private hospitals cannot expand readily to take care of them, whereas provision can be made for reserve facilities at city hospitals, for, as has been shown repeatedly in the past, in the face of a real epidemic governments can move swiftly to provide medical care. Third, unless or until the United States adopts a system of universal health care, public hospitals will be needed to care for those who fall between the cracks of the present hodgepodge of systems—those for whom laws and their implementation do not keep up with reality. Today's victims include those unprovided for by health insurance, Medicare, or Medicaid. In Florida and the Southwest their ranks are swelled by the large numbers of unregistered aliens. Finally, even if all patients could be assimilated into a system of private hospitals, it would be advisable to have at least one public-general hospital in a large city that would provide services that are vital to the public interest and serve as a stimulus or as a model for other hospitals to furnish needed services. And, if a system of universal health care were to be adopted, city hospitals would have the function of demonstrating how all hospitals should be responsive to the welfare of the general public.

The conclusion seems to be that it is not in the best interest of the people in metropolitan areas to close city hospitals. Furthermore, whether a city hospital is run by a city, county, or other local governmental authority, a public benefit corporation, or an arm of a state government such as a state university, it is important that it have adequate facilities and funding, professionalized administration, and close connections with academic health professions, particularly schools of medicine, nursing, and public health, and with the communities served.

By the mid-twentieth century city hospitals had become or were trying to become medical centers into which patients inserted themselves, were studied, treated, and turned out improved or recovered. The hospitals were ill-prepared for the contemporary period's emphasis on patients in their communities and on their continuous care. Some city hospitals still strained to become first-class centers for rendering inpatient care, while others also tried to relate to their communities by providing medical care that was continuous and expanded to comprehend the whole patient. Both objectives were thwarted by progressive fiscal problems of the cities. A movement arose to diminish the role of the locally owned public hospitals and to transfer them partly or wholly to private control. The relationships with communities, with the private hospital system, and with local, state, and federal governments are a Pandora's box of problems that has only just been opened. They await solution.

The history of city hospitals in the United States has revealed many changes: from custodial care in a poorhouse climate to scientific diagnosis and treatment in modern hospitals; from professional services donated by busy practitioners to full-time, highly organized staffs; from haphazard apprenticeships for professional workers to exacting educational programs; from subsistence on reluctant hand-outs from local governments to collections for services from patients, governments, and insurance carriers; from unreasoning fear that kept patients away to confidence in the hospitals' care that brings them in by the thousands; from patients who obeyed orders or got out to militant consumer groups who march and hire lawyers and sometimes participate in the hospital's planning. Yet, through these transformations several consistent threads weave in and out: public indifference and ignorance; manipulation for political ends; the parallel existence of voluntary hospitals that were sometimes rivals, sometimes allies, and sometimes merely fellow strugglers in a sea of despair; a multitude of patients who desperately needed the hospitals' services; and finally, a few dedicated workers in every rank who devoted their lives to the hospitals' service. New problems include the decaying central cities, the mounting costs of today's highly technical medical care, and the shifting sands of support from state and federal governments.

Despite their difficulties, city hospitals can survive if, in addition to rendering quality medical care of inpatients, they emphasize functions that meet other needs of people in their metropolitan areas—continuous care, preventive care, attention to environmental problems, and coordination with other agencies—and if they receive public support and enough resources to do the job. The vital question has always been and still remains: Does anybody care?

# Notes

## Introduction

Epigraph: *Aeneid,* 6.273.

1. Herbert E. Klarman, *Hospital Care in New York City: The Roles of Voluntary and Municipal Hospitals* (New York: Columbia University Press, 1963), p. 39; Report of the Commission on Public-General Hospitals, *The Future of the Public-General Hospital: An Agenda for Transition* (Chicago: Hospital Research and Educational Trust, 1978), pp. v–vi.

2. Commission on Public-General Hospitals, *Future of the Public-General Hospital,* p. 11.

3. Ibid., p. 37; Eli Ginzberg, *A Pattern for Hospital Care: Final Report of the New York State Hospital Study* (New York: Columbia University Press, 1949), pp. 144–145; U.S. Department of Commerce, Bureau of the Census, *Historical Statistics of the United States: Colonial Times to 1970,* 2 vols. (Washington, D.C.: U.S. Government Printing Office, 1975), 1:69–70.

4. Abraham Flexner, *Medical Education in the United States and Canada: A Report to the Carnegie Foundation for the Advancement of Teaching* (New York: Carnegie Foundation for the Advancement of Teaching, 1910).

## 1. Over the Hill from the Poorhouse

Epigraphs: Sir Thomas More, *Utopia,* ed. Edward Surtz (New Haven: Yale University Press, 1964), bk. 2, p. 78; Minutes of Managers of the Almshouse and House of Employment of Philadelphia, November 15, 1775, quoted in Robert J. Hunter, *The Origin of the Philadelphia General Hospital, Blockley Division* (Philadelphia: Rittenhouse Press, 1955), p. 15; circular letter, Boston, August 20, 1810, quoted in Thomas Francis Harrington, *The Harvard Medical School: A History, Narrative and Documentary, 1782–1905,* 3 vols. (New York: Lewis Publishing, 1905), 2:570.

1. William Hartson, "Care of the Sick Poor in England, 1572–1948," *Proceedings of the Royal Society of Medicine* 59 (June 1966):577–579; Saul Jarcho, "The Fate of British Traditions in the United States As Shown in Medical Education and in the Care of the Mentally Ill," *Bulletin of the New York Academy of Medicine* 52 (March-April 1976):431–434; George Rosen, *A History of Public Health* (New York: MD Publications, 1958), pp. 127–128; Charles Dickens, *Oliver Twist,* ch. 2.

2. Report of the Royal Commission, 1834, quoted in H.J. McCurrich, *The Treatment of the Sick Poor of this Country* (London: Oxford University Press, 1929), p. 29; Hartson, "Care of the Sick Poor in England," p. 581.

3. Jonas Hanway, *An Earnest Appeal for Mercy to the Children of the Poor* (1766), quoted in Karl de Schweinitz, *England's Road to Social Security: From the Statute of Laborers in 1349 to the Beveridge Report of 1942* (Philadelphia: University of Pennsylvania Press, 1943), p. 66.

4. De Schweinitz, *England's Road to Social Security,* p. 65.

5. McCurrich, *Treatment of the Sick Poor,* pp. 33–34.

6. Ibid., pp. 6–7; Brian Abel-Smith with Robert Pinker, *The Hospitals, 1800–1948: A Study in Social Administration in England and Wales* (London: Heinemann Educational Books, 1964), pp. 2–13, 49.

7. William Bennett Munro, *The Government of American Cities*, rev. ed. (New York: Macmillan, 1916), pp. 2n.1, 4, 4n.1,9.

8. Robert James Carlisle, *An Account of Bellevue Hospital with an Account of the Medical and Surgical Staff from 1736 to 1894* (New York: Society of the Alumni of Bellevue Hospital, 1893), p. 9; John Duffy, *A History of Public Health in New York City, 1625–1866* (New York: Russell Sage Foundation, 1968), pp. 67–68, 576.

9. Henry E. Sigerist, "An Outline of the Development of the Hospital," *Bulletin of the Institute of the History of Medicine* 4 (July 1936):573, 575; Rosen, *History of Public Health*, pp. 149–150. The quotation is from Massachusetts General Court, Committee on Pauper Laws, *Report of Committee to Whom Was Referred the Consideration of Pauper Laws of the Commonwealth* (1811), quoted in *The Heritage of American Social Work: Readings in Its Philosophical and Institutional Development*, ed. Ralph E. Pumphrey and Muriel W. Pumphrey (New York: Columbia University Press, 1961), p. 66.

10. Sigerist, "Outline of the Development of the Hospital," p. 580; David M. Schneider, *The History of Public Welfare in New York State, 1609–1866* (Chicago: University of Chicago Press, 1938), p. 21; Dieter Jetter, *Geschichte des Hospitals* (Wiesbaden: Franz Steiner, 1972), Band 3: *Nordamerika, 1600–1776: Kolonialzeit*, p. 72; N.I. Bowditch, *A History of the Massachusetts General Hospital*, 2d ed. (Boston: Trustees from the Bowditch Fund, 1872), p. 19.

11. Hunter, *Origin of the Philadelphia General Hospital*, pp. 11, 12, 13, 15, 20–23; the quotation is on p. 15.

12. Charles K. Mills, "The Philadelphia Almshouse and the Philadelphia Hospital from 1854 to 1908," in *Founders' Week Memorial Volume*, ed. Frederick P. Henry (Philadelphia, 1909), pp. 465–472; John Welsh Croskey, comp., *History of Blockley: A History of the Philadelphia General Hospital from Its Inception, 1731–1928* (Philadelphia: W.A. Davis, 1929), p. 15; Roland G. Curtin, "The Philadelphia General Hospital," in Henry, *Founders' Week Memorial Volume*, p. 424; D. Hayes Agnew, *Lecture on the Medical History of the Philadelphia Alms House Delivered at the Opening of the Clinical Lectures, October 15, 1862* (Philadelphia: Holland and Edgar, 1862), pp. 34–36.

13. Mills, "The Philadelphia Almshouse," p. 482; Joseph Chapman Doane, "A Brief History of the Philadelphia Hospital from 1908 to 1928," in Croskey, *History of Blockley*, pp. 112, 117–118; "Medical News," *Journal of the American Medical Association* 90 (March 10, 1928):780.

14. Schneider, *History of Public Welfare in New York State*, pp. 73, 75; Carlisle, *An Account of Bellevue Hospital*, pp. 4–6; John Starr, *Hospital City* (New York: Crown Publishers, 1957), pp. 8–9.

15. Carlisle, *An Account of Bellevue Hospital*, pp. 22–23, 33, 40.

16. Constance McLaughlin Green, *Washington: A History of the Capital, 1800–1950*, 2 vols. (Princeton, N.J.: Princeton University Press, 1962), 1:12–13.

17. Board of Health of the District of Columbia, *First Annual Report, 1872* (Washington, D.C.: Gibson Brothers, 1873), pp. 191–193. Constance McLaughlin Green states, however, that although the city took over the site in 1839, building of the asylum did not begin until 1843 (*Washington*, 1:136).

18. George M. Kober, *Charitable and Reformatory Institutions in the District of Columbia: History and Development of the Public Charitable and Reformatory Institutions and Agencies in the District of Columbia*, 69th Cong., 2d

sess., 1927, S. Doc. 207, pp. 19, 105–106; Green, *Washington*, 2:158, 167.

19. Douglas Carroll, "History of the Baltimore City Hospitals," *Maryland State Medical Journal* 15 (January 1966):88.

20. Ibid. (February 1966): 46–48; ibid. (August 1966): 69–71; ibid. (June 1966): 102; Thomas H. Wright, "Report to the Trustees on the State of the Medical Department of the Baltimore Alms-House Infirmary, for the Year Ending the 30th of April, 1831," *American Journal of the Medical Sciences* 10 (May 1832):60–61; George S. Mirick, "Medical Department of the Baltimore City Hospitals," *Maryland State Medical Journal* 4 (December 1955):759.

21. Circular letter, Boston, August 20, 1810, quoted in Harrington, *Harvard Medical School*, 2:570.

22. Journal of the Reverend Manasseh Cutler, vol. 1 (1787), quoted in Croskey, *History of Blockley*, pp. 14–15.

23. Fielding H. Garrison, *An Introduction to the History of Medicine* (Philadelphia: W.B. Saunders, 1914), pp. 309–310.

24. W.W. Gerhard, "Art. 1: On the Typhus Fever Which Occurred at Philadelphia in the Spring and Summer of 1836; Illustrated by Clinical Observations at the Philadelphia Hospital; Showing the Distinction between this Form of Disease and Dothinenteritis or the Typhoid Fever with Alteration of the Follicles of the Small Intestine," *American Journal of the Medical Sciences* 19 (1836):289; ibid. 20 (1837):289, both cited in Arthur L. Bloomfield, *A Bibliography of Internal Medicine* (Chicago: University of Chicago Press, 1958), pp. 6–7.

25. Hunter, *Origin of the Philadelphia General Hospital*, p. 13; D. Hayes Agnew, "The Medical History of the Philadelphia Almshouse," in Croskey, *History of Blockley*, pp. 30, 33; Garrison, *Introduction to the History of Medicine*, pp. 539–540.

26. Starr, *Hospital City*, pp. 13, 26, 49; William Frederick Norwood, *Medical Education in the United States before the Civil War* (Philadelphia: University of Pennsylvania Press, 1944), p. 53; Carroll, "History of the Baltimore City Hospitals," 15 (February 1966):46; Katherine A. Harvey, "Practicing Medicine at the Baltimore Almshouse, 1828–1850," *Maryland Historical Magazine* 74 (September 1979):224–225.

27. John Cary, *Joseph Warren: Physician, Politician, Patriot* (Urbana, Ill.: University of Illinois Press, 1961), p. 27; Rhoda Truax, *The Doctors Warren of Boston: First Family of Surgery* (Boston: Houghton Mifflin, 1968), pp. 42–43; Richard H. Shryock, *Eighteenth Century Medicine in America* (Worcester, Mass.: Davis Press, 1950), pp. 8–9; Norwood, *Medical Education in the United States*, pp. 9, 11, 429.

28. Leo James O'Hara, "An Emerging Profession: Philadelphia Medicine, 1860–1900" (Ph.D. diss., University of Pennsylvania, 1976), p. 341; Norwood, *Medical Education in the United States*, pp. 32–33; Richard Harrison Shryock, *The Development of Modern Medicine* (Philadelphia: University of Pennsylvania Press, 1936), p. 258.

29. Carlisle, *An Account of Bellevue Hospital*, pp. 36–45; Agnew, "Medical History of the Philadelphia Almshouse," p. 19.

30. Curtin, "The Philadelphia General Hospital," pp. 424, 454; Carroll, "History of the Baltimore City Hospitals," February 1966, p. 47; Carlisle, *An Account of Bellevue Hospital*, p. 33.

31. Harvey, "Practicing Medicine at the Baltimore Almshouse," p. 224; Carlisle, *An Account of Bellevue Hospital*, pp. 36–37, 43–48, 53.

32. Croskey, *History of Blockley*, pp. 31–32, 41–42.

33. Harvey, "Practicing Medicine at the Baltimore Almshouse," p. 230; Harry F. Dowling, *Fighting Infection: Conquests of the Twentieth Century* (Cambridge, Mass.: Harvard University Press, 1977), pp. 1–2, 6.

34. Quoted in Croskey, *History of Blockley*, pp. 14–15.

35. Charles Dickens, *Martin Chuzzlewit*, ch. 19; Charles Lawrence, *History of the Philadelphia Almshouses and Hospitals* (Philadelphia: Charles Lawrence, 1905), pp. 47–48.

36. Agnew, *Lecture on Medical History of the Philadelphia Alms House*, pp. 47–48.

37. Carlisle, *An Account of Bellevue Hospital*, pp. 38–39.

38. *New York Times*, April 25, 27, 1860.

39. Duffy, *Public Health in New York City*, p. 484.

40. Lawrence, *History of the Philadelphia Almshouses*, p. 32; Agnew, "Medical History of the Philadelphia Almshouse," p. 61; J. Chalmers Da Costa, "The Old Blockley Hospital: Its Characters and Characteristics," *Journal of the American Medical Association* 50 (April 11, 1908):1184.

41. Quoted in J. Thomas Scharf and Thompson Westcott, *History of Philadelphia, 1609–1884*, 3 vols. (Philadelphia: L.H. Everts, 1884), 2:1450.

42. "Isaac E. Taylor," in Samuel W. Francis, *Biographical Sketches of Distinguished Living New York Physicians* (New York: George P. Putnam and Son, 1867), pp. 140–141.

43. Curtin, "The Philadelphia General Hospital," pp. 429, 453–454; Agnew, "Medical History of the Philadelphia Almshouse," pp. 34–36, 44; Agnew, *Lecture on Medical History of the Philadelphia Alms House*, p. 20; Edward P. Cheyney, "The University of Pennsylvania, with Special Reference to the Medical and Allied Departments," in Henry, *Founders' Week Memorial Volume*, p. 224.

44. Agnew, *Lecture on Medical History of the Philadelphia Alms House*, pp. 21–22, 24–25; Agnew, "Medical History of the Philadelphia Almshouse," pp. 39–42.

45. Curtin, "The Philadelphia General Hospital," pp. 454–456; Mills, "The Philadelphia Almshouse," p. 509; Carlisle, *An Account of Bellevue Hospital*, pp. 47–49.

46. Claude Heaton, "Three Hundred Years of Medicine in New York City," *Bulletin of the History of Medicine* 32 (November-December 1958):522; Norwood, *Medical Education in the United States*, pp. 115, 135; James J. Walsh, *History of Medicine in New York*, 5 vols. (New York: National Americana Society, 1919), 2:474–475; Carlisle, *An Account of Bellevue Hospital*, pp. 36, 53, 58–59.

47. Carroll, "History of Baltimore City Hospitals," 15 (February 1966): 46–47.

## 2. Cities, Politicians, and City Hospitals

Epigraphs: Faun McKay Brodie, *Thomas Jefferson: An Intimate History* (New York: W.W. Norton, 1974), p. 72n; Andrew D. White, quoted in Richard Hofstadter, *The Age of Reform: From Bryan to F.D.R.* (New York: Alfred A. Knopf, 1955), p. 175; Major-General Hugh S. Cumming, "The Municipal Hospital of the Future and Its Relation to Public Health," *Medical Life* 40 (April 1933):159.

1. U.S. Department of Commerce, Bureau of the Census, *Historical Statistics of the United States: Colonial Times to 1970*, 2 vols. (Washington, D.C.: U.S. Government Printing Office, 1975), 1:11–12.

2. U.S. Department of Commerce, Bureau of Foreign and Domestic Commerce, *Statistical Abstract of the United States,1924* (Washington, D.C.: U.S. Government Printing Office, 1925), p. 7; U.S. Department of Commerce, Bureau of the Census, *Statistical Abstract of the United States, 1939* (Washington, D.C.: U.S. Government Printing Office, 1940), p. 11; U.S. Census Office, *Statistics of the United States...in 1860, Compiled from the Original Returns...of the Eighth Census*(1866; reprint ed., New York: Arno Press, 1976), pp. xviii, lviii–lviii.

3. W. Gill Wylie, *Hospitals: Their History, Organization, and Construction* (New York: D. Appleton, 1877), pp. 43, 205–206; E.H.L. Corwin, *The American Hospital* (New York: Commonwealth Fund, 1946), p. 181; Mary Risley, *House of Healing: The Story of the Hospital* (Garden City, N.Y.: Doubleday, 1961), pp. 222–223; S.S. Goldwater, "The United States Hospital Field," *National Hospital Record* 9 (April 1906):12.

4. Ignaz Philipp Semmelweis, "The Etiology, the Concept and the Prophylaxis of Childbed Fever," ed. and trans. Frank P. Murphy, *Medical Classics* 5 (January-April 1941):340–373, esp. pp. 356–357, 392–394; Joseph Lister, "On a New Method of Treating Compound Fractures, Abscess, etc., with Observations on the Conditions of Suppuration: Part I. On Compound Fracture," *Lancet* 1 (March 16, 1867):326–329; Joseph Lister, "The Address in Surgery," *British Medical Journal* 2 (August 26, 1871):225–233; Charles Singer and E. Ashworth Underwood, *A Short History of Medicine*, 2d ed. rev. (New York: Oxford University Press, 1962), pp. 354–358, 366–371; Jon Michael Kingsdale, "The Growth of Hospitals: An Economic History in Baltimore" (Ph.D. diss., University of Michigan, 1981), p. 92.

5. Another reason why the upper and middle classes began to patronize hospitals around the turn of the century may have been the decreasing size of living accommodations in cities as many small houses and apartment houses were erected. Kingsdale, "The Growth of Hospitals," pp. 103–108.

6. Constance Bell Webb, *A History of Contagious Disease Care in Chicago before the Great Fire* (Chicago: University of Chicago Press, 1940), pp. 26, 37, 78; David J. Davis, *History of Medical Practice in Illinois*, 2 vols. (Chicago: Illinois State Medical Society, 1927, 1955), vol. 2, *1850–1900*, p. 415; "Mercy Hospital," in Chicago Medical Society, *History of Medicine and Surgery and Physicians and Surgeons of Chicago* (Chicago: Biographical Publishing Company, 1922), pp. 235–236.

7. Fielding H. Garrison, *An Introduction to the History of Medicine* (Philadelphia: W.B. Saunders, 1914), p. 372; William E. Quine, "Early History of the Cook County Hospital to 1870," in Chicago Medical Society, *History of Medicine and Surgery*, pp. 257–260. For a good summary of homeopathy in the United States, see William G. Rothstein, *American Physicians in the Nineteenth Century: From Sects to Science* (Baltimore: Johns Hopkins University Press, 1972), pp. 152–174. For further discussion of the homeopathic branch of medicine in city hospitals, see Chapter 3.

8. Charles B. Johnson, *Growth of Cook County* (Chicago: Board of Commissioners of Cook County, Ill., 1960), pp. 172, 173–175; James Nevins Hyde, *Early Medical Chicago: An Historical Sketch* (Chicago: Fergus Printing Company, 1879), p. 42.

9. Private hospitals in Chicago at that time were Mercy, founded in 1850; St. Luke's, in 1863 or 1864; the Chicago Hospital for Women and Children, in 1865; and Alexian Brothers Hospital, in 1866. Bed capacities are not available for 1866, but in 1872 Mercy Hospital contained five hundred beds, St. Luke's fifty beds, the Hospital for Women and Children fifteen to twenty

beds, and Alexian Brothers two hundred beds. D.W. Graham, ed., *The Illinois State Medical Register for 1877–8* (Chicago: W.T. Keener, 1877), pp. 78–81; T. Davis Fitch and Norman Bridge, eds., *Chicago Medical Register and Directory, 1872–3* (Chicago: Hazlitt and Reed, 1872), pp. 193–194, 199, 204, 211.

10. William E. Quine, "Early History of the Cook County Hospital to 1870," in Chicago Medical Society, *History of Medicine and Surgery,* pp. 260, 261, 264.

11. H.M. Lyman, "A Bit of the History of Cook County Hospital," *Bulletin of the Society of Medical History of Chicago* 1 (October 1911):25–36; Graham, *The Illinois State Medical Register, 1877–78,* p. 79.

12. Henry Burdett, *Burdett's Hospitals and Charities, 1902* (London: Scientific Press, 1902), p. 840; E.J. Davis, "History, Development and Organization: Cook County Hospital," *Chicago Hospital Council Bulletin* 8 (September 1945):9.

13. Peter R. Knights, *The Plain People of Boston, 1830–1860: A Study in City Growth* (New York: Oxford University Press, 1971), p. 33; Oscar Handlin, *Boston's Immigrants: A Study in Acculturation,* rev. ed. (Cambridge, Mass.: Belknap Press of Harvard University Press, 1959), pp. 45–52; Edward Wakin, *The Immigrant Experience: Faith, Hope, and the Golden Door* (Huntington, Ind.: Our Sunday Visitor, 1977), p. 9; U.S. Census Office, *Statistics of the Eighth Census,* pp. xviii, lvii–lviii.

14. David W. Cheever et al., *A History of the Boston City Hospital from Its Founding until 1904* (Boston: Municipal Printing Office, 1906), p. 104.

15. Massachusetts General Hospital Trustees, printed letter, April 1, 1865, and "Report of a Committee on the Financial Condition of the Massachusetts General Hospital, February 16, 1865," cited by Morris J. Vogel, in "Patrons, Practitioners, and Patients: The Voluntary Hospital in Mid-Victorian Boston," in *Sickness and Health in America: Readings in the History of Medicine and Public Health,* ed. Judith Walzer Leavitt and Ronald L. Numbers (Madison, Wisc.: University of Wisconsin Press, 1978), pp. 176, 176n.

16. Boston Lying-In Hospital, *Forty-Second Annual Report* (Boston: Rand, Avery, 1875), title page; Morris J. Vogel, *The Invention of the Modern Hospital: Boston, 1870–1930* (Chicago: University of Chicago Press, 1980), p. 33.

17. Leonard K. Eaton, *New England Hospitals, 1790–1833* (Ann Arbor: University of Michigan Press, 1957), p. 54; N.I. Bowditch, *A History of the Massachusetts General Hospital,* 2d ed. (Boston: Trustees from the Bowditch Fund, 1872), pp. 36–574 passim; Morris J. Vogel, "Boston's Hospitals, 1870–1930: A Social History" (Ph.D. diss., University of Chicago, 1974), p. 51.

18. John Koren, *Boston, 1822 to 1922: The Story of Its Government and Principal Activities during One Hundred Years* (Boston: City Printing Department, 1923), pp. 138–142; Cheever et al., *History of the Boston City Hospital,* pp. 1–3; John J. Byrne, ed., *A History of the Boston City Hospital, 1905–1964* (Boston, 1964), pp. 1–6.

19. E.H.L. Corwin, "Rise of the Hospital Idea," *Bulletin of the New York Academy of Medicine* 9 (March 1933):115.

20. John Green, *City Hospitals* (Boston: Little, Brown, 1861), pp. 11–12.

21. Ibid., pp. 12, 13. The institution Green cited was the Massachusetts State Hospital at Rainsford Island near Boston.

22. City of Boston, *Proceedings at the Dedication of the Boston City Hospital* (Boston: J.E. Farwell, 1965), pp. 11, 12; the quotation is on p. 7.

23. Edward D. Churchill, ed., *To Work in the Vineyard of Surgery: The Reminiscences of J. Collins Warren (1842–1927)* (Cambridge, Mass.: Harvard University Press, 1958), pp. 138, 151n; "The Boston City's Fortieth Anniversary," *National Hospital Record* 7 (June 1904):28; Cheever et al., *History of the Boston City Hospital,* pp. 63–69.

24. D. Hayes Agnew, *Lecture on the Medical History of the Philadelphia Alms House Delivered at the Opening of the Clinical Lectures, October 15, 1862* (Philadelphia: Holland and Edgar, 1862), p. 31; Charles K. Mills, "The Philadelphia Almshouse and the Philadelphia Hospital from 1854 to 1908," in *History of Blockley: A History of the Philadelphia General Hospital from Its Inception, 1731–1928,* comp. John Welsh Croskey (Philadelphia: W.A. Davis, 1929), pp. 67, 81; Philadelphia Guardians for the Relief and Employment of the Poor, *Annual Statement, 1875* (Philadelphia: E.C. Markley and Son, 1876), p. 28.

25. Bellevue Hospital Training School for Nurses, Alumnae Association, *Bellevue: A Short History of Bellevue Hospital and of the Training Schools* (New York, 1915), pp. 2, 3 (picture); Robert James Carlisle, *An Account of Bellevue Hospital with an Account of the Medical and Surgical Staff from 1736 to 1894* (New York: Society of the Alumni of Bellevue Hospital, 1893), pp. 56, 63; Kingsdale, "The Growth of Hospitals," p. 95.

26. A hospital service consists of an attending physician or surgeon and the patients in his charge, together with other attending physicians and house officers who assist him in the care of those patients, and the medical students, if any, taught by the doctors on the service. One or more nursing units are assigned to a hospital service, depending on its size. Services generally correspond to the medical specialties (see Chapter 3).

27. Carlisle, *An Account of Bellevue Hospital,* pp. 70, 79, 100.

28. In addition to the almshouse hospital, Baltimore contained two medical school infirmaries at midcentury. Two hospitals opened in the 1850s and four in the 1860s, and by 1902 Baltimore had twenty private, nonprofit hospitals for the care of acutely ill patients, with a total capacity of 2,245 beds. Kingsdale, "The Growth of Hospitals," pp. 96, 180.

29. J. Thomas Scharf, *The Chronicles of Baltimore, Being a Complete History of "Baltimore Town" and Baltimore City from the Earliest Period to the Present Time* (1874; reprint ed., Port Washington, N.Y.: Kennikat Press, 1972), p. 75; Parker J. McMillin, "Baltimore City Hospitals," *Maryland State Medical Journal* 4 (December 1955):750; Douglas Carroll, "History of the Baltimore City Hospitals," ibid. 15 (June 1966):101–102; ibid. 15 (November 1966): 110.

30. W. Montague Cobb, "A Short History of Freedmen's Hospital," *Journal of the National Medical Association* 54 (May 1962):271–287; George M. Kober, *Charitable and Reformatory Institutions in the District of Columbia: History and Development of the Public Charitable and Reformatory Institutions and Agencies in the District of Columbia,* 69th Cong. 2d sess., 1927, S. Doc. 207, pp. 161–162.

31. Providence Alumnae Publication Committee, *Providence Centennial Book, 1861–1961* (Washington, D.C.: Providence Hospital, 1961), esp. pp. 18, 20; Columbia Hospital for Women and Lying-In Asylum, *Charter, By-Laws, &c., 1866* (Washington, D.C.: McGill and Witherow, 1866), pp. 6–8; ibid., *1876* (Washington, D.C.: U.S. Government Printing Office, 1876), p. 5; ibid., *1886* (Washington, D.C.: Gibson Bros., 1886), p. 7.

32. Kober, *Charitable and Reformatory Institutions,* pp. 49, 356 (the quotation is on pp. 108–109); Dorothy Jane Youtz, *The Capital City School*

*of Nursing, Formerly the Washington Training School for Nurses, 1877–1972,*
ed. Irene B. Page and Mary M. Goodreau (Washington, D.C.: Capital City
School of Nursing Alumni Association, 1975), p. 5; James A. Gannon,
"Recollections of Old Gallinger Hospital," *Medical Annals of the District of
Columbia* 20 (July 1951):391–393.

33. Albert E. Fossier, "The Charity Hospital of Louisiana," *New Orleans
Medical and Surgical Journal* 75 (May 1923):728–730; ibid. 76 (June
1923):791–798; ibid. 76 (July 1923):24–25; John Duffy, ed., *The Rudolph
Matas History of Medicine in Louisiana,* 2 vols. (Baton Rouge: Louisiana State
University Press, 1958, 1962), 1:247–255, 420–427.

34. Duffy, *History of Medicine in Louisiana,* 2: 208, 210, and 1:435 for the
quotation; Fossier, "Charity Hospital," 76 (August 1923):67–71.

35. *Report of the Board of Administrators of the Charity Hospital to the
General Assembly of Louisiana* (1879), quoted in Duffy, *History of Medicine in
Louisiana,* 2:501–502.

36. Fossier, "Charity Hospital," 76 (September 1923):132, 133, 137.

37. Commission on Hospital Care, *Hospital Care in the United States* (New
York: Commonwealth Fund, 1947), p. 445; James H. Rodabaugh and Mary
Jane Rodabaugh, *Nursing in Ohio: A History* (Columbus, Ohio: State Nurses
Association, 1951), p. 23; Roy R. Anderson and Maxine Beaton, "From Pest
Houses to Hospitals—Our Changing Hospital Scene," in *A Century of
Colorado Medicine, 1871–1971,* ed. Harvey T. Sethman (Denver: Colorado
Medical Society, 1971), p. 33; K.H. Van Norman, "Both Acute and Chronic
Patients Cared for in King County Hospitals," *Hospital Management* 47 (June
1939):25; Robert E. Tranquada and Robert F. Maronde, "The Hospital within
a Hospital: An Empirical Experiment in Health Care in a Major Metropolitan
Hospital," *Bulletin of the New York Academy of Medicine* 48 (April 1972):558.

38. Arthur Meier Schlesinger, *The Rise of the City, 1878–1898* (New
York: Macmillan, 1933), p. 79.

39. Howard P. Chudacoff, *The Evolution of American Urban Society*
(Englewood Cliffs, N.J.: Prentice-Hall, 1975), p. 126; Blake McKelvey, *The
Urbanization of America (1860–1915)* (New Brunswick, N.J.: Rutgers
University Press, 1963), pp. 86–88; Alexander B. Callow, Jr., ed., *The City
Boss in America: An Interpretive Reader* (New York: Oxford University Press,
1976), p. 5.

40. Carlisle, *An Account of Bellevue Hospital,* p. 57; *New York Times,*
October 2, 1895, January 3, 1896; City of New York, Bellevue and Allied
Hospitals, *First Annual Report, January 1, 1902 to December 31, 1902* (New
York: Martin B. Brown, 1903), p. 5; McMillin, "Baltimore City Hospitals," p.
751; Douglas Carroll, *Maryland State Medical Journal* "History of the
Baltimore City Hospitals," 15 (December 1966): 89–90.

41. Vogel, *Invention of the Modern Hospital,* pp. 29–31.

42. Ibid.; Morris J. Vogel, "Machine Politics and Medical Care: The City
Hospital at the Turn of the Century," in *The Therapeutic Revolution: Essays in
the Social History of American Medicine,* ed. Morris J. Vogel and Charles E.
Rosenberg (Philadelphia: University of Pennsylvania Press, 1979), pp.
159–175.

43. The quotation is in Jane Addams, *Twenty Years at Hull House* (1910;
Signet ed., New York: New American Library of World Literature, 1961), p.
127. See also Chudacoff, *Evolution of American Urban Society,* p. 125.

44. Callow, *The City Boss in America,* p. 5.

45. L.L. MacArthur, "Christian Fenger As I Knew Him," *Bulletin of the*

*Society of Medical History of Chicago* 1 (January 1923):10; J. Chalmers DaCosta, "The Old Blockley Hospital: Its Characters and Characteristics," in Croskey, *History of Blockley*, pp. 134–135.

46. Lloyd Wendt and Herman Kogan, *Lords of the Levee: The Story of Bathhouse John and Hinky Dink* (Indianapolis: Bobbs-Merrill, 1943), p. 28.

47. DaCosta, "Old Blockley Hospital," p. 134.

48. Brett Howard, *Boston: A Social History* (New York: Hawthorn Books, 1976), p. 68; Koren, *Boston, 1822 to 1922*, pp. 46–48; Leslie G. Ainley, *Boston Mahatma: The Public Career of Martin Lomasney* (Boston: W.M. Prendible, 1949); Vogel, *Invention of the Modern Hospital*, pp. 45, 46.

49. "Medical Notes," *Boston Medical and Surgical Journal* 148 (February 19, 1903):220.

50. Editorial, "Annual Report of the Boston City Hospital," *Boston Medical and Surgical Journal* 162 (January 20, 1910):84; Vogel, *Invention of the Modern Hospital*, pp. 45, 50–51, 57; "The Hospitals of Boston," *Modern Hospital* 1 (September 1913):53–54.

51. Charles S. Howell, "Association of Hospital Superintendents," *National Hospital Record* 3 (September 1899):1–3; Byrne, *History of the Boston City Hospital*, p. 390; Vogel, *Invention of the Modern Hospital*, pp. 37–41.

52. The quotation is in Roswell Park, "Hardships of an Intern," in *The Neoplasm*, ed. John James Mahoney, Ralph W. Avery, and Charles E. Boys (Chicago: G.J. Troutman Sons, 1902), p. 37; see also Quine, "Early History of the Cook County Hospital," pp. 263–264.

53. The quotation is from *Weekly Medical Review* (September 22, 1883), cited in Thomas Neville Bonner, *Medicine in Chicago, 1850–1950: A Chapter in the Social and Scientific Development of a City* (Madison, Wis.: American History Research Center, 1957), p. 163. See also Park, "Hardships of an Intern," p. 37; Loyal Davis, *J.B. Murphy: Stormy Petrel of Surgery* (New York: G.P. Putnam's Sons, 1938), pp. 121–122; Frank Billings, "History of the Cook County Hospital from 1876 to the Present Time," in Chicago Medical Society, *History of Medicine and Surgery*, pp. 265–266.

54. The quotation is from James B. Herrick, *Memories of Eighty Years* (Chicago: University of Chicago Press, 1949), pp. 70–71. See also Editorial, *Medical Standard* 4 (October 1888):115.

55. Charles K. Mills, "The Philadelphia Almshouse and the Philadelphia Hospital," pp. 73–74; Charles K. Mills, "The Philadelphia Almshouse and the Philadelphia Hospital from 1854 to 1908," in *Founders' Week Memorial Volume*, ed. Frederick P. Henry (Philadelphia, 1909), pp. 476–484; DaCosta, "Old Blockley Hospital," pp. 134, 138–140; Joseph Chapman Doane, "A Brief History of the Philadelphia Hospital from 1908 to 1928," in Croskey, *History of Blockley*, p. 111.

56. McMillin, "Baltimore City Hospitals," p. 751.

57. Constance McLaughlin Green, *Washington: A History of the Capital, 1800–1950*, 2 vols. (Princeton, N.J.: Princeton University Press, 1962), 1:313–362, 393–395, 2:71–72; Commissioners of the District of Columbia, *Summary Report of the Reorganization of the Government of the District of Columbia, July 1, 1952-September 30, 1953* (Washington, D.C.: Department of General Administration, Management Office, 1953), pp. 5–6; Kober, *Charitable and Reformatory Institutions*, pp. 9, 116.

58. Callow, *City Boss in America*, p. 3.

59. David Riesman, "How the New Blockley Came into Being," *Medical Life* 40 (March 1933):137–148; the quotation is on p. 142.

60. Ella M. Flick, *Beloved Crusader: Lawrence F. Flick* (Philadelphia: Dorrance, 1944), pp. 66, 73.

61. James Bryce, *The American Commonwealth*, rev. and abridged ed. (1906; reprint ed., New York: Macmillan, 1944), p. 429.

### 3. Practitioner-Doctors, City Hospitals, and Medical Schools

Epigraphs: Chaucer, *Canterbury Tales*, Prologue, line 308; Boston City Hospital, 1864, as described in David W. Cheever et al., *A History of the Boston City Hospital from Its Foundation until 1904* (Boston: Municipal Printing Office, 1906), p. 277.

1. John J. Byrne, ed., *A History of the Boston City Hospital, 1905–1964* (Boston, 1964), pp. 12–14; George W. Gay, "David Williams Cheever, A.M., M.D., LL.D.," *Boston Medical and Surgical Journal* 175 (July 20, 1916):71–77 (the quotations are on pp. 72, 73); "Dedication of the Cheever Amphitheater," ibid. 178 (April 4, 1918):449.

2. Mary M. Riddle, *Boston City Hospital Training School for Nurses: Historical Sketch* (Boston, 1928), p. 150; James Howard Means, *The Association of American Physicians: Its First Seventy-Five Years* (New York: McGraw-Hill, 1961), pp. 3, 6–7.

3. Byrne, *History of the Boston City Hospital*, p. 12; Morris J. Vogel, "Boston's Hospitals, 1870–1930: A Social History" (Ph.D. diss., University of Chicago, 1974), pp. 7–8, 11; Bard Gavin, ed., *Michael Freebern Gavin: A Biography* (Cambridge, Mass.: Riverside Press, 1915), p. 54; Morris J. Vogel, *The Invention of the Modern Hospital: Boston, 1870–1930* (Chicago: University of Chicago Press, 1980), p. 46; "Policy of Tufts College Medical School," *New England Journal of Medicine* 217 (September 30, 1937):564.

4. Elizabeth Jane Davis, "History, Development and Organization: Cook County Hospital," *Chicago Hospital Council Bulletin* 8 (September 1945):8; the quotation is from Matt. 26:52.

5. William E. Quine, "Early History of the Cook County Hospital to 1870," *Plexus* 16 (December 1910):588.

6. Karl A. Meyer, "Historical Background of Cook County Hospital," *Quarterly Bulletin of the Northwestern University Medical School* 23 (Fall 1949):272; Isaac A. Abt, "The Growth of Pediatrics in the Chicago Area," *Medical Clinics of North America* 30 (January 1946):10; W.B. Moulton, "The Merit System in Illinois," in *Public Welfare Administration in the United States: Select Documents*, ed. Sophonisba Breckinridge (New York: Johnson Reprint Corporation, 1970), pp. 450–453.

7. William Allen Pusey speculated that, as a newcomer to this country, Fenger assumed that paying for a staff appointment was a custom of the United States to which everyone was expected to conform. William Allen Pusey, "Highlights in the History of Chicago Medicine," *Bulletin of the Society of Medical History of Chicago* 5 (May 1940):182.

8. J. Christian Bay, *Dr. Christian Fenger: The Man and His Work*, address delivered at Northwestern University Medical School, November 4, 1940; "Great Names in Chicago Medicine: Christian Fenger," *Chicago Medicine* 63 (February 11, 1961):41–42; James B. Herrick, *Memories of Eighty Years* (Chicago: University of Chicago Press, 1949), pp. 168–175; L.L. McArthur, "Christian Fenger As I Knew Him," *Bulletin of the Society of Medical History of Chicago* 1 (January 1923):51–57; Christian Fenger, *Collected Works*, comp. Coleman G. Buford, ed. Ludwig Hektoen, 2 vols. (Philadelphia: W.B. Saunders, 1912), 1:1–5.

9. Loyal Davis, *J.B. Murphy: Stormy Petrel of Surgery* (New York: G.P. Putnam's Sons, 1938), p. 114; Herrick, *Memories of Eighty Years,* pp. 175–184 (the quotation is on p. 184). For the contributions of James Herrick, see Fredrick Arthur Willius and Thomas E. Keys, eds., *Cardiac Classics: A Collection of Classic Works on the Heart and Circulation, with Comprehensive Biographic Accounts of the Authors* (St. Louis: C.V. Mosby, 1941), pp. 815–816; James B. Herrick, "Peculiar Elongated and Sickle-Shaped Red Blood Corpuscles in a Case of Severe Anemia," *Archives of Internal Medicine* 6 (November 1910):517–521; James B. Herrick, "Clinical Features of Sudden Obstruction of the Coronary Arteries," *Journal of the American Medical Association* 59 (December 7, 1912):2015–2020.

10. Simon Flexner and James Thomas Flexner, *William Henry Welch and the Heroic Age of American Medicine* (New York: Viking Press, 1941), p. 68; W.G. MacCallum, *William Stewart Halsted: Surgeon* (Baltimore: Johns Hopkins Press, 1930), pp. 16–17.

11. Howard A. Kelly and Walter L. Burrage, *Dictionary of American Medical Biography* (New York: D. Appleton, 1928), p. 650.

12. Remarks of John Howland, "Memorial Meeting to Dr. Theodore Caldwell Janeway," *Bulletin of the Johns Hopkins Hospital* 29 (June 1918):143–144.

13. Harvey Cushing, *The Life of Sir William Osler,* 2 vols. (Oxford: Clarendon Press, 1925), 1:251–253, 275, 285, 440.

14. Herrick, *Memories of Eighty Years,* p. 68; H.M. Lyman, "A Bit of the History of Cook County Hospital," *Bulletin of the Society of Medical History of Chicago* 1 (October 1911):31.

15. MacCallum, *William Stewart Halsted,* p. 41; McArthur, "Christian Fenger As I Knew Him," p. 54.

16. MacCallum, *William Stewart Halsted,* p. 19.

17. Cushing, *Life of Osler,* 1:237–238.

18. Louis J. Mitchell, "The Cook County Hospital," *North American Practitioner* 1 (August 1889):398–400.

19. "Bellevue Hospital: History and Early Organization," *New York Journal of Medicine,* n.s. 16 (May 1856):392; *New York Times,* November 8, 1891.

20. William Shaine Middleton, "Clinical Teaching in the Philadelphia Almshouse and Hospitals (Part II)," *Medical Life* 40 (May 1933):221; J.H. Capps, "Dr. Frank Billings," *Quarterly Bulletin of the Northwestern University Medical School* 30 (Winter 1956):378; Stephen Smith, "Report on the General Hospitals of the State of New York," in New York State Board of Charities, *Twenty-Ninth Annual Report* (New York, 1896), p. 280; "Chicago News Items," *National Hospital Record* 9 (September 1905):37; Sumner Koch, "A Tribute to Walter Herman Nadler, April 30, 1889-September 13, 1952," *Quarterly Bulletin of the Northwestern University Medical School* 26 (Winter 1952):405–407; Edwin Frederick Hirsch, *Frank Billings* (Chicago: University of Chicago, 1966), p. 133.

21. The quotation is from Editorial, "Cook County Hospital," *Medical Standard* 24 (May 1901):236; see also "From the Editor's Note Book," ibid., 24 (June 1901):333.

22. Hermann Hagedorn, *Leonard Wood: A Biography,* 2 vols. (New York: Harper and Brothers, 1931), 1:38; George H.M. Rowe, *Observations on Hospital Organization* (1902), p. 13, reprinted from *National Hospital Record* 6 (December 1902):3–10.

23. Julia C. Lathrop, "The Cook County Charities," in Residents of Hull-

House, *Hull-House Maps and Papers* (New York: Thomas Y. Crowell, 1895), pp. 155–156.

24. *New York Times,* January 16, 1900.

25. D. Hayes Agnew, "The Medical History of the Philadelphia Almshouse," in *History of Blockley: A History of the Philadelphia General Hospital from Its Inception, 1731–1928,* comp. John Welsh Croskey (Philadelphia: F.A. Davis, 1929), p. 27; Cheever et al., *History of the Boston City Hospital,* p. 8.

26. Cheever et al., *History of the Boston City Hospital,* pp. 154, 157; Edgar M. Bick, "American Orthopedic Surgery: The First 200 Years," *Bulletin of the New York Academy of Medicine* 52 (March-April 1976):303; *New York Times,* March 12, 1875; Charles K. Mills, "The Philadelphia Almshouse and the Philadelphia Hospital from 1854 to 1908," in Croskey, *History of Blockley,* p. 95.

27. Knowlton E. Barber, "History of Urology in Chicago, Part III," *Chicago Medicine* 63 (May 27, 1961):35; Adrian W. Zorgniotti, "The Creation of the American Urologist, 1902–1912," *Bulletin of the New York Academy of Medicine* 52 (March-April 1976):287; Byrne, *History of the Boston City Hospital,* pp. 232–255.

28. Issac A. Abt, "A Survey of Pediatrics during the Past 100 Years," *Illinois Medical Journal* 77 (May 1940):491–492; Byrne, *History of the Boston City Hospital,* p. 151; Harold E. Harrison, "The Pediatric Service [at Baltimore City Hospitals]," *Maryland State Medical Journal* 4 (December 1955):777.

29. Robert James Carlisle, *An Account of Bellevue Hospital with an Account of the Medical and Surgical Staff from 1736 to 1894* (New York: Society of the Alumni of Bellevue Hospital, 1893), pp. 63–64; *New York Times,* March 12, 1875; Mills, "The Philadelphia Almshouse and the Philadelphia Hospital," pp. 80–81, 85.

30. The quotation is from Henry M. Lyman, "A Bit of the History of Cook County Hospital," *Bulletin of the Society of Medical History of Chicago* 1 (October 1941):25–36. See also William E. Quine, "Early History of the Cook County Hospital to 1870," in Chicago Medical Society, *History of Medicine and Surgery and Physicians and Surgeons of Chicago* (Chicago: Biographical Publishing Corporation, 1922), p. 261; Davis, "Cook County Hospital," p. 9; William A. Mann, "Great Names in Chicago Medicine: The History of Ophthalmology in Chicago, Part II," *Chicago Medicine* 63 (April 22, 1961):38; Malcolm T. MacEachern, "Summary Report of Findings and Recommendations from the Survey of the Cook County Hospital and Cook County School of Nursing" (mimeographed, Chicago, 1932), p. 11.

31. William G. Rothstein, *American Physicians in the Nineteenth Century: From Sects to Science* (Baltimore: Johns Hopkins University Press, 1972), pp. 209–211.

32. Board of Health of the District of Columbia, *First Annual Report, 1872* (Washington, D.C.: Gibson Brothers, 1873), p. 193.

33. Carlisle, *An Account of Bellevue Hospital,* pp. 12–15; Board of Health of the District of Columbia, *First Annual Report,* pp. 191–193; Byrne, *History of the Boston City Hospital,* p. 25.

34. Cheever et al., *History of the Boston City Hospital,* pp. 131–139. Figures on diphtheria are from William Osler, *The Principles and Practice of Medicine,* 4th ed. (New York: D. Appleton, 1902), pp. 156–157.

35. S.S. Woody, "Municipal Hospitals for Contagious Diseases," *American Journal of Public Health* 2 (September 1912):726–732; James J. Walsh,

*History of Medicine in New York*, 5 vols. (New York: National Americana Society, 1919), 3:730–733; William H. Welch, "The Municipal Hospital for Contagious and Infectious Diseases," in *Founders' Week Memorial Volume*, ed. Frederick P. Henry (Philadelphia, 1909), pp. 521, 526; William Travis Howard, Jr., *Public Health Administration and the Natural History of Disease in Baltimore, Maryland, 1787–1920* (Washington, D.C.: Carnegie Institution, 1924), p. 19; Huntington Williams, "The Origins of the Baltimore City Medical Care Program, 1776–1948," *Baltimore Health News* 26 (July 1949):138*n;* Commissioners of the District of Columbia, *Report of the Health Officer, 1900* (Washington, D.C.: U.S. Government Printing Office, 1900), pp. 16–17.

Not until 1935 did Washington abolish the system of farming out contagious diseases to private hospitals and open a modern unit at the city hospital. Baltimore closed its separate hospital for communicable diseases in 1950 and moved these patients to the city hospital. Washington, D.C., Commissioners, *Report of the Government of the District of Columbia for the Year Ended June 30, 1935* (Washington, D.C.: U.S. Government Printing Office, 1936), p. 128; Dorothy Jane Youtz, *The Capital City School of Nursing, Formerly the Washington Training School for Nurses, 1877–1972*, ed. Irene B. Page and Mary M. Goodreau (Washington, D.C.: Capital City School of Nursing Alumni Association, 1975), pp. 91–93; Douglas Carroll, "History of the Baltimore City Hospitals," *Maryland State Medical Journal* 15 (October 1966):89; Thomas B. Turner, *Heritage of Excellence: The Johns Hopkins Medical Institutions, 1914–1947* (Baltimore: Johns Hopkins University Press, 1974), p. 307.

36. Editorial, "Results of Isolation of Fever," *American Medical Times* 9 (August 1864):66–67.

37. Cushing, *Life of Osler*, 1:253.

38. Flexner and Flexner, *William Henry Welch*, pp. 111–137; *Biographical Sketches and Letters of T. Mitchell Prudden, M.D.* (New Haven: Yale University Press, 1927), p. 54; About Pollack, " A Brief History of the Pathology Laboratory of the Baltimore City Hospitals," *Maryland State Medical Journal* 4 (December 1955):781.

39. Albert Nelson Marquis, ed., *Who's Who in America*, (Chicago: A.N. Marquis, 1912–1913), p. 1574; Pollack, "The Pathology Laboratory of the Baltimore City Hospitals," p. 781.

40. Henry K. Beecher and Mark D. Altschule, *Medicine at Harvard: The First Three Hundred Years* (Hanover, N.H.: University Press of New England, 1977), p. 115; Timothy Leary, "Frank Burr Mallory and the Pathological Department of the Boston City Hospital," *American Journal of Pathology* 9, supp. (1933):659–672; Byrne, *History of the Boston City Hospital*, p. 309.

41. Quine, "Early History of the Cook County Hospital," pp. 583–584; William Harvey King, *History of Homeopathy and Its Institutions in America*, 4 vols. (New York: Lewis Publishing, 1905), 1:352; Thomas Neville Bonner, *Medicine in Chicago, 1850–1950: A Chapter in the Social and Scientific Development of a City* (Madison, Wis.: American History Research Center, 1957), p. 41; "Cook County Hospital," *National Hospital and Sanitarium Record* 1 (January 1898):4.

42. Boston City Hospital, *Action of the Trustees upon the Petitions for the Introduction of Homeopathic Treatment...* (Boston: Rockwell and Churchill, 1886), pp. 3–9.

43. Agnew, "Medical History of the Philadelphia Almshouse," p. 42.

44. Kate Campbell Hurd-Mead, *Medical Women of America: A Short History*

*of the Pioneer Medical Women of America and a Few of Their Colleagues in England* (New York: Froben Press, 1933), pp. 52–53; Kelly and Burrage, *Dictionary of American Medical Biography,* pp. 331–332, 1163; Croskey, *History of Blockley,* pp. 258–259; Esther Pohl Lovejoy, *Women Doctors of the World* (New York: Macmillan, 1957), p. 93; Mills, "The Philadelphia Almshouse and Philadelphia Hospital," p. 99; Harold J. Abrahams, *The Extinct Medical Schools of Baltimore, Maryland* (Baltimore: Maryland Historical Society, 1969), p. 72; "Cook County Hospital, Chicago," *National Hospital Record* 2 (April 1899):27; Helen MacMurchy, "Hospital Appointments: Are They Open to Women?" *New York Medical Journal* 73 (April 27, 1901):714–716; Mary Roth Walsh, *"Doctors Wanted: No Women Need Apply"—Sexual Barriers in the Medical Profession, 1835–1975* (New Haven, Conn.: Yale University Press, 1977), pp. 223, 224; City of Boston, *Remarks on the Subject of Admitting Female Students to Clinical Lectures at the City Hospital* (Boston, 1871).

45. Arba Nelson Waterman, *Historical Review of Chicago and Cook County and Selected Biography,* 3 vols. (Chicago: Lewis Publishing, 1908), 1:307, 310–311, 316–318; *The Pulse of Rush Medical College* (Chicago: Middle Class of Rush Medical College, 1895), p. 22; *New York Times,* May 13, 1882; Beecher and Altschule, *Medicine at Harvard,* p. 95; Chicago Medical Society, *History of Medicine and Surgery,* p. 194; *Chicago Tribune,* July 14, 1895.

46. Edward F. Dolan, Jr., *Vanquishing Yellow Fever: Walter Reed* (Chicago: Brittanica Books, 1962), pp. 50–51.

47. Abraham Flexner, *Medical Education in the United States and Canada: A Report to the Carnegie Foundation for the Advancement of Teaching* (New York: Carnegie Foundation for the Advancement of Teaching, 1910), pp. 232, 240–241, 323. Abraham Flexner was a teacher and author in education who was employed by the Carnegie Foundation to make a study of American medical schools. His report resulted in the improvement or abandonment of the many inferior medical schools and set the pattern for medical education for the next half century.

48. William F. Norwood, "Education in the United States before 1900," in *History of Medical Education,* ed. C.D. O'Malley (Berkeley: University of California Press, 1970), p. 477; U.S. Department of Commerce, Bureau of the Census, *Historical Statistics of the United States: Colonial Times to 1970,* 2 vols. (Washington, D.C.: U.S. Government Printing Office, 1975), 1:8.

49. Frank Billings later recalled how he was "passed back" by upperclassmen to the last row from a front seat, where he had näively sat. "Frank Billings," *Proceedings of the Institute of Medicine of Chicago* 26 (September 1966):109.

50. Albert E. Fossier, "History of Medical Education in New Orleans from Its Birth to the Civil War," *Annals of Medical History,* n.s. 6 (July 1934):347; John Eric Erichsen, "Impressions of American Surgery," *Lancet* 2 (November 21, 1874):718; Carlisle, *An Account of Bellevue Hospital,* p. 87; Chicago Medical Society, *History of Medicine and Surgery in Chicago,* p. 194.

51. For instance, in the New York Hospital in 1886 "instruction in the amphitheatre [was] almost entirely confined to surgeons, one physician only giving such instruction, and that irregularly." Boston City Hospital, *Action of the Trustees,* p. 15. See also Gert Brieger, ed., *Medical America in the Nineteenth Century* (Baltimore: Johns Hopkins Press, 1972), p. 37.

52. John Shrady, *The College of Physicians and Surgeons, New York, and Its Founders, Officers, Instructors, Benefactors and Alumni,* 2 vols. (New York: Lewis Publishing, 1904), 1:300.

53. George M. Gould, *The Jefferson Medical College of Philadelphia ...1826–1904: A History* (New York: Lewis Publishing, 1904), pp. 183–184; J.W. Holland, "The Jefferson Medical College of Philadelphia," in Henry, *Founders' Week Memorial Volume,* p. 272; James B. Herrick, "The Ward Clinic," *National Hospital Record* 6 (February 1903):57–61.

54. Cushing, *Life of Osler,* 1:284–285.

55. David Riesman, "Clinical Teaching in America, with Some Remarks on Early Medical Schools," *Transactions and Studies of the College of Physicians of Philadelphia,* 4th ser., 7 (April 1939):109; quotation of Dr. Lewis A. Corner from James A. Harrar, *The Story of the Lying-In Hospital of New York* (New York: Society of the Lying-In Hospital, 1938), p. 75.

56. Herrick, *Memories of Eighty Years,* p. 79.

57. Boston City Hospital, *Action of the Trustees,* pp. 22–23; Thomas Francis Harrington, *The Harvard Medical School: A History, Narrative and Documentary,* ed. James Gregory Mumford, 3 vols. (New York: Lewis Publishing, 1905), 2:1115; Flexner, *Medical Education in the United States and Canada,* p. 218; Leo James O'Hara, "An Emerging Profession: Philadelphia Medicine, 1860–1900" (Ph.D. diss., University of Pennsylvania, 1976), pp. 136–137.

58. Fossier, "History of Medical Education in New Orleans," n.s. 6 (July 1934):347; ibid. (September 1934):432–434, 439–440.

59. The quotation is from Herbert L. Burrell and J. Bapst Blake, "The Teaching of Surgery at the Boston City Hospital," *Medical and Surgical Reports of the Boston City Hospital* 11 (1900):116. See also Alan M. Chesney, *The Johns Hopkins Hospital and the Johns Hopkins University School of Medicine: A Chronicle,* 3 vols. (Baltimore: Johns Hopkins Press, 1943, 1958, 1963), vol. 2, *1893–1905,* pp. 130–135; Harvard University Faculty of Medicine, *The Harvard Medical School, 1782–1906* (Boston, 1906), pp. 8, 78; Beecher and Altschule, *Medicine at Harvard,* pp. 110–111; Henry A. Christian, "Medical Teaching at Harvard and the Opportunity It Offers the Young Graduate for Medical Training," *Boston Medical and Surgical Journal* 157 (July 18, 1907): 67–69; Joseph C. Aub and Ruth K. Hapgood, *Pioneer in Modern Medicine: David Linn Edsall of Harvard* (Boston: Harvard Medical Alumni Association, 1970), p. 160.

60. Flexner, *Medical Education in the United States and Canada,* p. 115.

61. Richard Harrison Shryock, *The Development of Modern Medicine: An Interpretation of the Social and Scientific Factors Involved* (New York: Alfred A. Knopf, 1947), pp. 273–335 passim; Harry F. Dowling, *Fighting Infection: Conquests of the Twentieth Century* (Cambridge, Mass.: Harvard University Press, 1977).

## 4. Comfort and Care: The Ongoing Struggle

Epigraph: W.H. Rideing, "Hospital Life in New York," *Harper's New Monthly Magazine* (1878), quoted in *Medical America in the Nineteenth Century: Readings from the Literature,* ed. Gert H. Brieger (Baltimore: Johns Hopkins Press, 1972), p. 246.

1. Depending on which version of the story one reads, he had either been burned by a lamp or had been bruised and cut falling against a washbasin. John Tasker Howard, *America's Troubadour* (New York: Thomas Y. Crowell, 1934), pp. 333–342; Harvey Gaul, *The Minstrel of the Alleghenies* (Pittsburgh: Friends of Harvey Gaul, 1934), pp. 82–83; Harold Vincent

Milligan, *Stephen Collins Foster: A Biography of America's Folk-Song Composer* (New York: G. Schirmer, 1920), pp. 102, 106.

2. For diagnoses in 1867, see Trustees of the Boston City Hospital, *Fourth Annual Report* (1867), p. 27, cited in Morris J. Vogel, *The Invention of the Modern Hospital: Boston, 1870–1930* (Chicago: University of Chicago Press, 1980), p. 60. The quotation is from Samuel B. Woodward, "Address at Annual Meeting of the Boston City Hospital Alumni Association," *New England Journal of Medicine* 215 (July 2, 1936):27.

3. Charles Rosenberg, "And Heal the Sick: The Hospital and the Patient in 19th Century America," *Journal of Social History* 10 (January 1977):431, 432.

4. Gladys Sellew and C.J. Nuesse, *A History of Nursing* (St. Louis: C.V. Mosby, 1946), p. 281; Elizabeth M. Jamieson and Mary F. Sewall, *Trends in Nursing History*, 4th ed. (Philadelphia: W.B. Saunders, 1954), pp. 272–273; Richard Harrison Shryock, "Nursing Emerges As a Profession: The American Experience," in *Sickness and Health in America: Readings in the History of Medicine and Public Health*, ed. Judith Walzer Leavitt and Ronald L. Numbers (Madison, Wis.: University of Wisconsin Press, 1978), p. 204.

5. Harvey Cushing, *The Life of Sir William Osler*, 2 vols. (Oxford: Clarendon Press, 1925), 1:290; Arthur Ames Bliss, *Blockley Days: Memoirs and Impressions of a Resident Physician, 1883–1884* (Springfield, Mass.: Springfield Printing & Binding, 1916), pp. 19–20. Osler was referring to Thomas J. Owens, head nurse of the men's medical floor. Joseph Chapman Doane, "A Brief History of the Philadelphia General Hospital from 1908 to 1928," in *History of Blockley: A History of the Philadelphia General Hospital from Its Inception, 1731–1928*, comp. John Welsh Croskey (Philadelphia: W.A. Davis, 1929), p. 109.

6. The quotation is from Linda Richards, *Reminiscences of America's First Trained Nurse* (Boston: Whitcomb and Barrows, 1911), pp. 6–7. See also M. Adelaide Nutting and Lavinia L. Dock, *A History of Nursing*, 2 vols. (New York: G.P. Putnam's Sons, 1907), 2:352n, 419–421, 422.

7. The quotation is from Bliss, *Blockley Days*, p. 20. See also ibid., pp. 24–26; Shryock, "Nursing Emerges As a Profession," p. 204.

8. Shryock, "Nursing Emerges As a Profession," pp. 204–205; Richard Harrison Shryock, *The Development of Modern Medicine: An Interpretation of the Social and Scientific Factors Involved* (New York: Alfred A. Knopf, 1947), pp. 343–344.

9. The quotation is from John Boyle O'Reilly, *Selected Poems* (1904; 2d ed., New York: H.M. Caldwell, 1913), p. 23. See also Alexander B. Callow, "The Crusade against the Tweed Ring," in *American Urban History: An Interpretive Reader with Commentaries*, ed. Alexander B. Callow (New York: Oxford University Press, 1969), pp. 330–348; Nutting and Dock, *History of Nursing*, 2:358–362, 375–376.

10. Elizabeth C. Hobson, *Recollections of a Happy Life*, n.d., quoted in Dorothy Giles, *A Candle in Her Hand: A Story of the Nursing Schools of Bellevue Hospital* (New York: G.P. Putnam's Sons, 1949), p. 70.

11. W. Gill Wylie, *Hospitals: Their History, Organization, and Construction* (New York: D. Appleton, 1877), pp. 3–4.

12. A few schools had opened in the United States in hospitals for women beginning in 1858, but their training programs were considerably shorter and more limited than in the schools of the Nightingale type. Vern L. Bullough and Bonnie Bullough, *The Emergence of Modern Nursing*, 2d ed. (New York: Macmillan, 1969), pp. 123–124; Philip A. Kalisch and Beatrice J. Kalisch,

*The Advance of American Nursing* (Boston: Little, Brown, 1978), pp. 79, 86–88.

13. Robert James Carlisle, *An Account of Bellevue Hospital with an Account of the Medical and Surgical Staff from 1736 to 1894* (New York: Society of the Alumni of Bellevue Hospital, 1893), p. 79; John Starr, *Hospital City* (New York: Crown Publishers, 1957), p. 105; Bullough and Bullough, *The Emergence of Modern Nursing*, pp. 123–124; Kalisch and Kalisch, *The Advance of American Nursing*, pp. 79, 86–88; W.G. Thompson, *Training-Schools for Nurses with Notes on Twenty-two Schools* (New York: G.P. Putnam's Sons, 1883), pp. 31–35, 38; "New York City Training School for Nurses," *National Hospital Record* 6 (August 1902):7; "The Boston City Hospital Training School for Nurses," ibid., p. 48.

14. Grace Fay Schryver, *A History of the Illinois Training School for Nurses, 1880–1929* (Chicago: Board of Directors of the Illinois Training School for Nurses, 1930), pp. 71, 74, 82–83; the quotation is on p. 62.

15. Ibid., pp. 54–55; *Chicago Daily News*, December 9, 1895; *Chicago Tribune*, December 10, 1895, December 11, 1895; *Inter-Ocean* (Chicago), January 15, 1893; Schryver, *A History of the Illinois Training School for Nurses*, pp. 165–176; James A. Hamilton and Associates, "Study of Nursing in the Cook County Hospital, Chicago, Illinois" (mimeographed, Chicago, 1950), pp. II–36–II–37, available at Cook County Hospital Archives, Chicago.

16. Mary M. Riddle, *Boston City Hospital Training School for Nurses: Historical Sketch* (Boston, 1928), p. 18; John J. Byrne, ed., *A History of the Boston City Hospital, 1905–1964* (Boston, 1964), p. 362.

17. Joseph McFarland, "The History of Nursing at the Blockley Hospital," *Medical Life* 40 (April 1933):182–184, 188; Charles K. Mills, "The Philadelphia Almshouse and the Philadelphia Hospital from 1854 to 1908," in Croskey, *History of Blockley*, pp. 91–92.

18. Albert E. Fossier, "The Charity Hospital of Louisiana," *New Orleans Medical and Surgical Journal* 76 (September 1923):136–138; John Duffy, ed., *The Rudolph Matas History of Medicine in Louisiana*, 2 vols. (Baton Rouge: Louisiana State University Press, 1958, 1962), 2:503–507.

19. Dorothy Jane Youtz, *The Capital City School of Nursing, Formerly the Washington Training School for Nurses, 1877–1972*, ed. Irene B. Page and Mary M. Goodreau (Washington, D.C.: Capital City School of Nursing Alumni Association, 1975), pp. 13, 15, 19, 24, 36; Superintendent of Charities, Washington, D.C., *Report on Charitable and Reformatory Institutions, 1898* (Washington, D.C.: U.S. Government Printing Office, 1898), pp. 5, 53.

20. Douglas Carroll, "History of the Baltimore City Hospitals," *Maryland State Medical Journal* 15 (August 1966): 70; Baltimore Department of Charities and Correction, Supervisors of City Charities, *Annual Report, 1915* (Baltimore, 1916), p. 26; Baltimore Supervisors of City Charities, Minutes, April 1, 1925, Maryland Historical Society, Baltimore.

21. Thompson, *Training-Schools for Nurses*, pp. 25–51; Agnes E. Pavey, *The Story of the Growth of Nursing* (London: Faber and Faber, 1959), pp. 379, 381; Cushing, *Life of Osler*, 1:295.

22. The quotation is from McFarland, "History of Nursing at Blockley Hospital," p. 186. See also Barton Cooke Hirst, "Some Reminiscences of the Philadelphia Hospital As a Prospective Interne, and As a Member of the Staff for 20 Years," *Medical Life* 40 (May 1933):235.

23. Byrne, *A History of the Boston City Hospital*, pp. 22, 52; Shryock, "Nursing Emerges As a Profession," p. 206.

24. *New York Times,* May 17, 1893; May 4, 1900; December 23, 1900; December 30, 1900; January 10, 1901.

25. Boston City Hospital, Trustees Records, August 20, 1884, quoted in Vogel, *Invention of the Modern Hospital,* p. 40.

26. Editorial, *New York Times,* October 5, 1882; Frederick P. Henry, "Scurvy," *Philadelphia Hospital Reports* 1 (1890): 125–151; Cook County Hospital, *Annual Report of the Warden for the Year Ending August 31, 1884* (Chicago: J.M.W. Jones Stationery and Printing, 1885), p. 14; U.S. Department of Commerce, Bureau of the Census, *Historical Statistics of the United States: Colonial Times to 1970,* 2 vols. (Washington, D.C.: U.S. Government Printing Office, 1975), 1:211, 212.

27. Cook County Hospital, *Annual Report, 1884,* p. 14; Trustees of the Boston City Hospital, *Nineteenth Annual Report* (Boston: Rockwell and Churchill, 1884), p. 10; U.S. Department of Commerce, Bureau of the Census, *Historical Statistics,* 1:211, 212; Cook County Hospital, *Annual Reports for Fiscal Year 1915* (Chicago, n.d.), pp. 13–14; Baltimore Department of Charities and Correction, *Annual Report, 1915,* p. 19; Board of Charities of the District of Columbia, *Annual Report, 1901* (Washington, D.C.: U.S. Government Printing Office, 1901), pp. 24–25; Rosenberg, "And Heal the Sick," pp. 429–431, 438–439.

28. Cook County Hospital, *Annual Report, 1908* (Chicago: Henry O. Shepard, 1909), p. 17; Cook County Hospital, *Annual Report, 1915,* p. 7; Baltimore Supervisors of City Charities, Minutes, April 7, 1919.

29. Trustees of the Boston City Hospital, *Seventh Annual Report, 1871* (Boston: Alfred Mudge and Son, 1871), pp. 7, 8; Trustees of the Boston City Hospital, *Forty-seventh Annual Report* (Boston, 1911), pp. 42, 43, 49, 89, 90; William E. Quine, "Early History of the Cook County Hospital to 1870," *Plexus* 16 (December 1910):586; Elizabeth Jane Davis, "History, Development and Organization: Cook County Hospital," *Chicago Hospital Council Bulletin* 8 (September 1943):8; Leslie B. Arey, *Northwestern University Medical School, 1859–1959: A Pioneer in Education Reform* (Evanston, Ill.: Northwestern University Medical School, 1959), p. 294; Henry Burdett, *Burdett's Hospitals and Charities* (London: Scientific Press, 1913), p. 767.

30. Isaac Ray, *What Shall Philadelphia Do for Its Paupers?,* paper read before the Social Science Association of Philadelphia, March 27, 1873, n.p., pp. 6, 17; Fossier, "Charity Hospital of Louisiana," 76 (October 1923):193.

31. Clara A. Goldwater and C.-E. A. Winslow, "A Biographical Note," in S.S. Goldwater, *On Hospitals* (New York: Macmillan, 1949), p. xxiii.

32. Homer Folks, "The City's Health—Public Hospitals, with Special Reference to New York City," *Municipal Affairs* 2 (June 1898):272–273, 275–276.

33. Charles Singer and E. Ashworth Underwood, *A Short History of Medicine,* 2d ed. (New York: Oxford University Press, 1962), pp. 374–375; Byrne, *History of Boston City Hospital,* p. 33; Max Ritvo, "X-Ray BCH First: Established Here Soon after Discovery in '95," *BCH Progress Notes,* January 1958, n.p.; George E. Pfahler, "Our Blockley Heritage," *Bulletin of the Society of Medical History of Chicago* 5 (May 1939):148; Edward S. Blaine, "The X-Ray Department of Cook County Hospital," *Modern Hospital* 8 (May 1917):321–328.

34. Vogel, *Invention of the Modern Hospital,* pp. 23–25; the quotation is on p. 24.

35. Morris J. Vogel, "Patrons, Practitioners, and Patients: The Voluntary Hospital in Mid-Victorian Boston," in Leavitt and Numbers, *Sickness and*

*Health in America*, p. 180; Thomas Neville Bonner, *Medicine in Chicago, 1850–1950: A Chapter in the Social and Scientific Development of a City* (Madison, Wis.: American History of Research Center, 1957), p. 156.

36. The quotation is from Jane Addams, "The Layman's View of Hospital Work among the Poor," *Transactions of the American Hospital Association* 9 (1907):57–58. See also Editorial, "Medical Social Work at the Boston City Hospital," *Boston Medical and Surgical Journal* 174 (February 17, 1916): 246; Joseph Hirsch and Beka Doherty, *The First Hundred Years of the Mount Sinai Hospital of New York* (New York: Random House, 1952), p. 139; Malcolm T. MacEachern, *Hospital Organization and Management*, 3rd ed., rev. (Chicago: Physicians' Record Company, 1957), p. 694; Alan M. Chesney, *The Johns Hopkins Hospital and the Johns Hopkins University School of Medicine: A Chronicle*, 3 vols. (Baltimore: Johns Hopkins Press, 1943, 1958, 1963), vol. 3, *1905–1914*, p. 57.

37. Bellevue Hospital, Training School for Nurses, Alumnae Association, *Bellevue: A Short History of Bellevue Hospital and of the Training Schools* (New York, 1915), p. 5; "A New Feature at Bellevue Hospital," *National Hospital Record* 10 (January 1907):64; Schryver, *Illinois Training School for Nurses*, pp. 143–144; "Practical Work of Social Service Department," *Hospital Management* 2 (September 1916):9–11.

38. Editorial, "Medical Social Work at the Boston City Hospital"; Miscellany: Medico-Social Work at the Boston City Hospital," *Boston Medical and Surgical Journal* 174 (February 17, 1916):255–256; Doane, "History of the Philadelphia General Hospital from 1908 to 1928," p. 120; Elizabeth Wisner, *Public Welfare Administration in Louisiana* (Chicago: University of Chicago Press, 1930), pp. 67, 199.

39. The quotation is from Boston City Hospital, *Proceedings at the Dedication of the City Hospital* (Boston: J.E. Farwell, 1865), p. 7. See also Vogel, *Invention of the Modern Hospital*, pp. 70, 71. Regarding venereal disease, the qualifying word *acute* may have been added because the moralists thought that those with a more chronic type of disease would have had time to repent; but more likely it was because doctors knew by the mid-nineteenth century that the chronic complications of venereal diseases could involve many different organs and could often be indistinguishable from other diseases.

40. David W. Cheever et al., *A History of the Boston City Hospital from Its Foundation until 1904* (Boston: Municipal Printing Office, 1906), p. 8; conversations with Maxwell Finland and William S. Jordan, Washington, D.C., May 10, 1979; Trustees of the Boston City Hospital, *Seventh Annual Report*, pp. 8, 9; Vogel, *Invention of the Modern Hospital*, pp. 109–110, 112–114. For figures on private patients, see City of Boston, *Report on Fees for Medical Treatment at the City Hospital*, City Document no. 35 (Boston, 1868), quoted in ibid., p. 106.

41. The quotations are from Trustees of the Boston City Hospital, *Fourteenth Annual Report* (Boston: Rockwell and Churchill, 1878), p. 6; and Editorial, "Report of the Massachusetts General and Boston City Hospitals," *Boston Medical and Surgical Journal* 156 (May 9, 1907):613–614. See also Cheever et al., *History of Boston City Hospital*, p. 167; Vogel, *Invention of the Modern Hospital*, pp. 118–119.

42. Board of Charities of the District of Columbia, *Annual Report* (Washington, D.C.: U.S. Government Printing Office, 1924), p. 5; Youtz, *Capital City School of Nursing*, p. 108.

43. *New York Times*, August 7, 1887.

## 5. Cities and Their Hospitals: An Uneasy Equilibrium

Epigraphs: Description of Jefferson Davis Hospital, Houston, in Jan de Hartog, *The Hospital* (New York: Atheneum, 1964), pp. 5–6; Lewis Thomas, "Public Teaching Hospitals Face Life-Death Issues in Changing Times," *Modern Hospital* 109 (October 1967):95.

1. U.S. Department of Commerce, Bureau of the Census, *Historical Statistics of the United States: Colonial Times to 1970.* 2 vols. (Washington, D.C.: U.S. Government Printing Office, 1975), 1:8, 11, 12.

2. See Table 6 for sources of population figures.

3. U.S. Department of Commerce, Bureau of the Census, *Historical Statistics.* 1:79; *Hospitals.* Guide Issue 25 (June 1951):1–159.

4. The quotation is from Rupert Norton, "Municipal Hospitals and Their Relation to the Community," *Modern Hospital* 1 (September 1913):43. See also Blake McKelvey, *The Urbanization of America (1860–1915)* (New Brunswick, N.J.: Rutgers University Press, 1963), pp. 86–88; Seth Low, "An American View of Municipal Government in the United States," in James Bryce, *The American Commonwealth.* rev. ed. (1906; abridged ed., New York: Macmillan, 1944), pp. 441–442.

5. John Koren, *Boston. 1822 to 1922: The Story of Its Government and Principal Activities during One Hundred Years* (Boston: City Printing Department, 1923), pp. 12–15; Low, "An American View of Municipal Government," pp. 441–442; Brett Howard, *Boston: A Social History* (New York: Hawthorn Books, 1976), p. 68; "Address of Dr. H.S. Rowen of the Board of Trustees of the Boston City Hospital before the Alumni Association," *Boston Medical and Surgical Journal* 192 (May 7, 1925):924–926.

6. James Michael Curley, *I'd Do It Again: A Record of All My Uproarious Years* (Englewood Cliffs, N.J.: Prentice-Hall, 1957), pp. 30–31.

7. Editorial, "The Health Commissionership of Boston," *Boston Medical and Surgical Journal* 193 (November 19, 1925):985–986.

8. Conversation with Robert Williams, Seattle, Wash., October 17, 1977; conversation with Maxwell Finland, Washington, D.C., May 10, 1980; Francis Russell, *The Great Interlude: Neglected Events and Persons from the First World War to the Depression* (New York: McGraw-Hill, 1964), pp. 211–212; conversation with William Jordan, Washington, D.C., May 10, 1979.

9. Editorial, "Pettifogging Petty Politics," *New England Journal of Medicine* 218 (February 17, 1938):315; conversation with Maxwell Finland, May 10, 1980.

10. This entry is dated January 10, 1934. The diary is available in the Harry Filmore Dowling Papers, History of Medicine Division, National Library of Medicine, Bethesda, Md.

11. "New Building for Boston City Hospital," *Boston Medical and Surgical Journal* 194 (January 7, 1926):35; John J. Dowling, "Plans and Progress of the Boston City Hospital," *New England Journal of Medicine* 198 (May 24, 1928):723–724; Hazel W. Bridges, "Mothers and Their Babies Cared for at the Boston City Hospital in 1929," ibid. 204 (April 23, 1931):869–879; American Medical Association, Council on Medical Education and Hospitals, "Hospital Service in the United States, 1931: The 1930 Census of Hospitals," *Journal of the American Medical Association* 96 (March 28, 1931):1045; "Obstetrics at the Boston City Hospital," *New England Journal of Medicine* 204 (April 23, 1931):886.

12. "Additions to the Boston City Hospital," *New England Journal of*

*Medicine* 202 (January 23, 1930):172; "New Buildings and Construction," *Bulletin of the American Hospital Association* 4 (February 1930):12; ibid. 6 (August 1932):51; "Addition to the Boston City Hospital," *New England Journal of Medicine* 207 (September 29, 1932):589; *Hospitals,* Guide Issue 30 (August 1, 1956):156; Cook County Board of Commissioners, *Annual Report, 1955* (Chicago: Black and Associates, 1955), pp. 34, 128.

13. Frances Burns, "Why Boston Can't Afford to Lose the City Hospital," series of eleven articles in *Boston Globe,* November 16–26, 1958, reprinted by city of Boston, December 1958; James V. Sacchetti to House Officers at Boston City Hospital, June 23, 1958, mimeographed, Boston City Hospital Library; Editorial, "The Boston City Hospital—Its Future Prospects," *New England Journal of Medicine* 262 (March 31, 1960):681–682; Robert H. Hamlin, Arnold I. Kisch, and H. Jack Geiger, "Administrative Reorganization of Municipal Health Services—The Boston Experience," ibid. 273 (July 1, 1963): 26–29; Robert H. Hamlin, Arnold I. Kisch, and H. Jack Geiger, "Report on the City Hospital and Boston Health Department to Mayor John F. Collins and the Board of Trustees of Boston City Hospital, March 29, 1965" (mimeographed); "Trustees' Report to Mayor Holds Out Hope of Continued Life for Boston City Hospital," *Hospitals* 43 (February 1, 1969):119–120; James B. Sacchetti, "108-Year-Old Boston City Hospital Undergoing Extensive Modernization," *Hospital Topics* 50 (June 1972):52, 72; "Boston Moves to Update Health," *Medical World News* 6 (May 28, 1965):71–73.

14. Norton, "Municipal Hospitals and Their Relation to the Community," p. 43.

15. Frank Billings, "History of the Cook County Hospital from 1876 to the Present Time," in Chicago Medical Society, *History of Medicine and Surgery and Physicians and Surgeons of Chicago* (Chicago: Biographical Publishing Corporation, 1922), p. 265; Karl A. Meyer, "Historical Background of Cook County Hospital," *Quarterly Bulletin of Northwestern University Medical School* 23 (Fall 1949):273; "Children's Unit To Be Added to Cook County," *Modern Hospital* 26 (February 1926):149.

16. M.T. MacEachern and E.W. Williamson, "Report of Survey: Cook County Hospital, Chicago, Illinois" (mimeographed; Chicago: Hospital Standardization Department, American College of Surgeons, 1927), pp. 1, 5–6, 9–10, 11, 44.

17. Elizabeth Jane Davis, "History, Development and Organization: Cook County Hospital," *Chicago Hospital Council Bulletin* 8 (September 1945):10, 11, 21; Meyer, "Historical Background of Cook County Hospital,"pp. 273–274.

18. According to Richard B. Ogilvie, former governor of Illinois and previously a member of the Cook County Board of Commissioners, "the lack of public concern coupled with the institution's employment needs have long made the public hospital a patronage dream." "Symposium Takes Up Problems, Approaches Used to Maintain Public Hospital Services," *Hospitals* 44 (March 1, 1970):105.

19. Charles B. Johnson, *Growth of Cook County* (Chicago: Board of Commissioners of Cook County, Ill., 1960), pp. 166–168, 170; Editorial, "The Future of Cook County Hospital," *Chicago Medicine* 67 (May 30, 1964):437.

20. Malcolm T. MacEachern, "Summary Report of Findings and Recommendations from the Survey of the Cook County Hospital and Cook County

School of Nursing" (mimeographed, Chicago, 1932), p. 7. Pierre DeVise, "The Problem Isn't How Little We Spend on the Poor, But How Badly We Spend," *Modern Hospital* 114 (May 1970): 86.

21. *Who's Who in America, 1950–51* (Chicago: A.N. Marquis, 1950), p. 1881; Johnson, *Growth of Cook County,* pp. 216–224; Harold Levine, "Chicago: Revolution within the Establishment," *New Physician* 19 (October 1970):838.

22. Frank Smith, "Misery Harbor: What Goes on behind County Hospital Doors," *Chicago Times,* series of articles, February-March 1936; American Medical Association Study of Medical Care, "Report of Committee on Medical Economics of the Chicago Medical Society," *Journal of the American Medical Association* 113 (July 8, 1939):151; Advisory Committee to the Citizens' Advisory Council of Cook County Hospital, "Medical Services in Cook County Hospital," *Proceedings of the Institute of Medicine of Chicago* 12 (April 15, 1939):345–346; "County Hospital Off AMA and ACS Approved Lists," *Chicago Hospital Council Bulletin* 1 (October 1938):5; Johnson, *Growth of Cook County,* pp. 216–224.

23. "Appointment of Brig. Gen. McCloskey," *Chicago Hospital Council Bulletin* 1 (September 1938):6; "Report of Committee on Survey of Cook County Hospital," ibid., p. 12.

24. Citizens' Committee to Study Cook County Hospital, "Final Report" (mimeographed, Chicago, 1964), p. 7; Editorial, "Commissioners Appoint Administrator for Cook County Hospital," *Proceedings of the Institute of Medicine of Chicago* 25 (July 1965):273.

25. The quotation is from "Cook County Hospital," *Resident Physician* 1 (September 1955):55. See also Robert J. Freeark, "The Plight of the Public Hospital. Case Study: Chicago," *Hospitals* 44 (July 1, 1970):57.

26. The quotations are from Cook County Board of Commissioners, *Annual Message, 1939* (Chicago: Leo P. Dywer, 1939), p. 91; and Theodore H. White, *In Search of History: A Personal Adventure* (New York: Harper and Row, 1978), p. 537. See also *Chicago Tribune Magazine,* July 26, 1970.

27. DeVise, "The Problem Isn't How Little We Spend," p. 86.

28. James A. Gannon, "Recollections of Old Gallinger Hospital," *Medical Annals of the District of Columbia* 20 (July 1951):391–392; George M. Kober, *Charitable and Reformatory Institutions in the District of Columbia: History and Development of the Public Charitable and Reformatory Institutions and Agencies in the District of Columbia,* 69th Cong., 2d sess., 1927, S. Doc. 207, p. 144.

29. The District of Columbia got its first mayor, Walter Washington, in 1967. The reorganized executive branch of the city government, which replaced the cumbersome three-man board of commissioners, included a mayor (at first called first commissioner), named by the President of the United States, and a nine-man city council. "Two Firsts for Washington," *Time,* September 15, 1967, pp. 23–24. The mayor became an elected official in 1975. Congress remained the legislative branch of the city's government.

30. U.S. Congress, Senate, *Conditions at Gallinger Municipal Hospital: Hearings before a Subcommittee of the Committee on the District of Columbia,* 78th Cong., 1st sess., September 27-October 1, 1943, p. 153.

31. Gannon, "Recollections of Gallinger Hospital," p. 391.

32. *Summary Report of the Reorganization of the Government of the District of Columbia, July 1, 1952-September 30, 1953* (Washington, D.C.: Department of General Administration Management Office, 1953), pp. 5–6; Constance McLaughlin Green, *Washington: A History of the Capital, 1800–1950,* 2 vols.

(Princeton, N.J.: Princeton University Press, 1962), 1:393–395; U.S. Congress, House, *Report of the Commission on the Organization of the Government of the District of Columbia*, 92nd Cong., 2d sess., 1972, H. Doc. 92–317, 4 vols., vol. 1, *Summary*, pp. 235–248; Board of Charities of the District of Columbia, *Annual Report, 1915* (Washington, D.C.: U.S. Government Printing Office, 1915), p. 35; Washington, D.C., Commissioners, *Report of the Government of the District of Columbia for the Year Ended June 30, 1935* (Washington, D.C.: U.S. Government Printing Office, 1936), p. 136.

33. Quoted in Constance McLaughlin Green, *The Secret City: A History of Race Relations in the Nation's Capital* (Princeton, N.J.: Princeton University Press, 1967), p. 247.

34. The Washington Asylum received appropriations of a similar amount during this period, but the majority of these funds were spent on maintenance of the workhouse and almshouse, which accommodated five to six times as many inmates as did the hospital. Board of Charities of the District of Columbia, *Annual Reports, 1901–1906.*

35. Kober, *Charitable and Reformatory Institutions*, p. 124.

36. *Washington Star*, May 1, 1916, March 7, 1917; *Washington Post*, December 18, 1916; Green, *Washington*, 2:450; Board of Public Welfare of the District of Columbia, *Annual Report of Charitable and Correctional Institutions and Agencies, 1928* (Washington, D.C.: U.S. Government Printing Office, 1928), p. 60. See Table 7 for sources of hospital statistics.

37. *Washington Star*, September 11, 1929, November 26, 1937, November 27, 1937; *Washington Times*, September 22, December 18, 1937; *Washington Herald*, July 10, 1937, November 30, 1937; *Washington Post*, February 17, 1936, March 1, 1936, November 28, 1937; U.S. Congress, Senate, *Congressional Record*, 78th Cong., 1st sess., 1943, p. 7584; U.S. Congress, Senate, *Conditions at Gallinger Municipal Hospital*, pp. 3–85.

38. U.S. Congress, Senate, Committee on the District of Columbia, Subcommittee on Public Health, Education, Welfare, and Safety, *District of Columbia Hospital and Health Facility Construction and Modernization*, 90th Cong., 1st sess., August 21, 1967, p. 88; *Washington Star*, April 28, 1949, March 15, 1950, February 13, 1962; *Washington Post*, February 14, 1950, June 9, 1959, January 8, 1962; *Washington News*, March 5, 1956, March 8, 1956, November 18, 1958.

39. Charles Hirschfeld, *Baltimore, 1870–1900: Studies in Social History* (Baltimore: Johns Hopkins University Press, 1941), p. 153; William Travis Howard, *Public Health Administration and the Natural History of Disease in Baltimore, Maryland, 1797–1920* (Washington, D.C.: Carnegie Institution of Washington, 1924), p. 19; Edmund G. Beacham, "History of Tuberculosis Division, Baltimore City Hospitals," *Maryland State Medical Journal* 4 (December 1955):771; George S. Mirick, "Medical Department of the Baltimore City Hospitals," ibid., p. 759.

40. The quotation is from the *Baltimore Sun*, June 28, 1931. See also ibid., December 18, 1920, January 3, 1921; Editorial, *Baltimore Evening Sun*, January 3, 1921; *A Surgeon's Life: The Autobiography of J.M.T. Finney* (New York: G.P. Putnam's Sons, 1940), pp. 316–325; Minutes of the Board of Supervisors of City Charities, Baltimore, Maryland, December 14, 1921, February 17, 1922, April 7, 1922, July 7, 1922, December 7, 1923, April 4, 1924, August 12, 1924, Maryland Historical Society, Baltimore; Baltimore Department of Charities and Correction, Supervisors of City Charities, *Annual Report, 1925* (Baltimore, 1925), p. 27; ibid., *1920*, p. 23; ibid., *1923*, p. 23.

41. Minutes of the Board of Supervisors of City Charities, Baltimore, Maryland, January 5, 1923, January 21, 1923, April 4, 1924; *Baltimore Sun*, June 28, 1931, September 7, 1933; Baltimore Department of Charities and Correction, Supervisors of City Charities, *Annual Report, 1925*, p. 12; ibid., *1911*, p. 7. For sources of cost figures, see Table 11, and Bellevue and Allied Hospitals, *Annual Report, 1925* (New York, 1925), p. 41.

42. The quotation is from Harvey Cushing, *The Life of Sir William Osler*, 2 vols. (Oxford: Clarendon Press, 1925), 1:381. See also *Baltimore Sun*, June 28, 1931.

43. John J. Byrne, ed., *A History of the Boston City Hospital, 1905–1964* (Boston, 1964), p. 33; "News of the Month," *Medical Standard* 25 (March 1902): 163; Cook County Commissioners, *Annual Report, 1923* (Chicago, n.d.), p. 12; Boston City Hospital, *Fifty-seventh Annual Report* (Boston, 1921), p. 98.

44. Mirick, "Medical Department of Baltimore City Hospitals," p. 760; Alan M. Chesney, *The Hospitals of Baltimore and the Indigent Sick* (Baltimore, 1932), p. 12.

45. This was confirmed as recently as 1979 at Harlem Hospital in New York City. Since this institution had no scanner for computed tomography examinations, arrangements had been made for examinations to be performed on its patients at two other hospitals. Yet, among patients recommended for this procedure, only 252 of 1,528 patients without trauma and a mere six of 342 with trauma received the examination. Reasons for the failure to perform the test were unavailability of the facilities in the assisting hospital at the time the test was needed or unavailability of transportation. It is obvious that hospitals with major medical commitments need major equipment and cannot subsist upon facilities in other hospitals, regardless of good intentions. John C.M. Brust, P.C. Taylor Dickinson, and Edward B. Healton, "Failure of CT Sharing in a Large Municipal Hospital," *New England Journal of Medicine* 304 (June 4, 1981): 1388–1393.

46. *Baltimore Sun*, September 7, 1933.

47. "In Memoriam," in Baltimore Department of Public Welfare, *Third Annual Report* (Baltimore: King Brothers, 1937), p. 211; *Baltimore Sun*, June 28, 1931; Douglas Carroll, "History of the Baltimore City Hospitals," *Maryland State Medical Journal* 15 (September 1966): 105–106.

48. Parker J. McMillin, "Baltimore City Hospitals," *Maryland State Medical Journal* 4 (December 1955): 751; "Parker J. McMillin," *The Record* (Baltimore) 1 (November-December 1958): 1–2; Carroll, "History of Baltimore City Hospitals," pp. 105–106; Harry F. Dowling, *Fighting Infection: Conquests of the Twentieth Century* (Cambridge, Mass.: Harvard University Press, 1977), pp. 158–173; Beacham, "History of Tuberculosis Division," p. 772; Mirick, "Medical Department of Baltimore City Hospitals," p. 761.

49. E.G. Beacham, D.G. Carroll, and F.G. Hubbard, "Baltimore City Hospitals: Toward a Third Century of Progress," *Maryland State Medical Journal* 22 (November 1973): 45, 46; Beacham, "History of Tuberculosis Division," p. 772; conversation with Charles C.J. Carpenter, Washington, D.C., May 10, 1979; conversation with Edyth Schoenrich, Baltimore, Md., November 17, 1980.

50. "Baltimore City Hospitals: Third Century of Service," printed pamphlet, n.p., n.d.; Beacham, Carroll, and Hubbard, "Baltimore City Hospitals," pp. 46–48.

51. Beacham, Carroll, and Hubbard, "Baltimore City Hospitals," p. 45; *Hospitals*, Guide Issue 40 (August 1, 1966): 103.

52. John J. Byrne, ed., *A History of the Boston City Hospital, 1905–1964* (Boston, 1964), p. 390; Minutes of the Board of Supervisors of City Charities, Baltimore, Maryland, March 18, 1921.

53. Morris J. Vogel, *The Invention of the Modern Hospital: Boston, 1870–1930* (Chicago: University of Chicago Press, 1980), p. 69.

54. James B. Crooks, *Politics and Progress: The Rise of Urban Progressivism in Baltimore, 1895 to 1911* (Baton Rouge, La.: Louisiana State University Press, 1968); Baltimore Department of Public Welfare, *First Annual Report* (Baltimore: Daily Record Press, 1935), p. 143; McMillin, "Baltimore City Hospitals," p. 751.

55. Through the 1860s all eleven trustees appointed had names of English origin, and in the remainder of the nineteenth century sixteen had English-sounding names, four Irish, and three had names of other derivation. In the twentieth century Irish names surpassed English twelve to eleven, and three were of other origin. Byrne, *History of the Boston City Hospital,* pp. 388–389.

56. Joseph Dever, *Cushing of Boston: A Candid Portrait* (Boston: Bruce Humphries, 1965), pp. 231–237; John Henry Cutler, *Cardinal Cushing of Boston* (New York: Hawthorn Books, 1970), pp. 109–110; conversation with Maxwell Finland, Boston, Mass., April 20, 1978.

57. Commissioners of the District of Columbia, *Annual Report of the Government of the District of Columbia, 1938* (Washington, D.C.: U.S. Government Printing Office, 1938), p. 165.

58. It was not uncommon for Karl Meyer, while presiding over meetings of the Hektoen Institute trustees, to request advice on hospital matters unrelated to the institute. In this way he could obtain opinions without obligating himself to follow them or even to report back.

59. Ethel H. Davis, *The Hektoen Institute for Medical Research* (Chicago: Kuttner and Kuttner, n.d.), p. 5.

60. Clara A. Goldwater and C.-E.A. Winslow, "A Biographical Note," in S.S. Goldwater, *On Hospitals* (New York: Macmillan, 1949), p. xiv; T. Harry Williams, *Huey Long* (New York: Bantam Books, 1970), pp. 306–308; Hermann B. Deutsch, *The Huey Long Murder Case* (Garden City, N.Y.: Doubleday, 1963), pp. 19, 540.

61. Williams, *Huey Long,* pp. 542–546, 918–919, 919n; Rudolph Matas, "Dr. Charles C. Bass, Dean: An Appreciation," *New Orleans Medical and Surgical Journal* 92 (April 1940):549–550.

62. Len O'Connor, *Clout: Mayor Daley and His City* (New York: Avon Books, 1975), pp. 173, 174.

63. *Washington Post* and *Washington Star,* August 4-October 3, 1929; *Washington Herald,* July 10, 1931; *Washington Star,* November 27, 1937, February 26, 1939, September 12, 1943, March 5, 1948, March 28, 1957; *Washington Post,* September 12, 1956; *Washington Daily News,* November 18, 1958; U.S. Senate, *Conditions at Gallinger Municipal Hospital.*

64. Later, Malcolm T. MacEachern, in *Hospital Organization and Management,* 3d ed. rev. (Chicago: Physicians' Record Company, 1957), p. 24, summarized the objects of the programs of hospital standardization as follows: "The focus of all effort in Hospital Standardization was the patient. It was with the object of giving him the best professional, scientific, and humanitarian care that the entire program was conducted and stipulations such as the following required of the approved hospital: that it have an organized, competent, and ethical medical staff; that the staff hold regular conferences for review of the clinical work;...that accurate and complete clinical records be kept of all patients treated; and that adequate diagnostic and therapeutic

facilities...be provided. The growth of this movement is remarkable when it is remembered that acceptance of the standard was entirely voluntary."

65. Franklin H. Martin, *Fifty Years of Medicine and Surgery: An Autobiographical Sketch* (Chicago: Surgical Publishing Company, 1934), pp. 337–338. In 1951 the Joint Commission on Accreditation of Hospitals was organized to continue the accreditation processes begun by the American College of Surgeons. By 1976 the sponsoring organizations were the American Medical Association, American Hospital Association, American College of Surgeons, American College of Physicians, American Association of Homes for the Aging, and the American Nursing Home Association. Rockwell Schultz and Alton C. Johnson, *Management of Hospitals* (New York: McGraw-Hill, 1976), p. 259.

66. American College of Surgeons, *Manual of Hospital Standardization: History, Development, and Progress of Hospital Standardization, Detailed Explanation of the Minimum Requirements* (Chicago, 1940), p. 7. The quotation is from J.D. Porterfield, "Hospital Accreditation—Past, Present and Future," *American Journal of Hospital Pharmacy* 27 (April 1970):315.

67. Lists of hospitals existing during this period are somewhat inconsistent in their use of the terms *public* and *private,* leading the reader to conclude that *public* may refer to a nonprofit institution, whether government owned or not. Thus, it is impossible to obtain an exact count of city and county hospitals in 1920; the estimate used here is based on the list of hospitals in *American Medical Directory,* 7th ed. (Chicago: American Medical Association, 1921).

68. "General Hospitals of 100 or More Beds: Report for 1920," *Bulletin of the American College of Surgeons* 5 (January 1921):10–13.

69. Davis, "History of Cook County Hospital," p. 10; "Cook County Hospital Faces Crisis over Accreditation and Hiring of an Administrator," *Modern Hospital* 102 (May 1964): 176; "One-Year Accreditations Given Two Major Municipal Hospitals," *Hospitals* 44 (May 16, 1970): 119–120; "Briefs of the News," *Modern Hospital* 116 (February 1971):155.

### 6. The Development of University Services

Epigraphs: Maxwell Finland, "Teaching and Learning in Government-Supported Hospitals," *Pharos* 39 (October 1976):132; "Remarks by Edward M. Dempsey at the Reopening of the Thorndike Memorial Laboratory, December 10, 1964," in *Training of the Physician: Papers Presented as Part of the Celebration of the Centennial of the Boston City Hospital and the Fortieth Anniversary of the Thorndike Memorial Laboratory* (Boston, 1964), p. 123.

1. David W. Cheever et al., *A History of the Boston City Hospital from Its Foundation until 1904* (Boston: Municipal Printing Office, 1906), pp. 156, 156n, 283; John J. Byrne, ed., *A History of the Boston City Hospital, 1905–1964* (Boston, 1964), pp. 178, 181; Morris J. Vogel, *The Invention of the Modern Hospital: Boston, 1870–1930* (Chicago: University of Chicago Press, 1980), p. 81.

2. Harry F. Dowling, *Fighting Infection: Conquests of the Twentieth Century* (Cambridge, Mass.: Harvard University Press, 1977).

3. The quotation of President Eliot is from Samuel Eliot Morison, *The Development of Harvard University since the Inauguration of President Eliot, 1869–1929* (Cambridge, Mass.: Harvard University Press, 1930), p. 565. See also ibid., p. 573; F.B. Mallory, "The Present Needs of the Harvard Medical School," *Science* 24 (September 14, 1906) :334–338; Harvard

University School of Medicine, "Minutes and Records of the Faculty of Medicine," November 7, 1890, December n.d., 1890, May 16, 1891, December 5, 1903, vol. 5, pp. 24, 28, 52, 230–31, Francis A. Countway Library of Medicine, Boston, Mass.; Frederic A. Washburn, *The Massachusetts General Hospital: Its Development, 1900–1935* (Boston: Houghton Mifflin, 1939), p. 91; Joseph C. Aub and Ruth K. Hapgood, *Pioneer in Modern Medicine: David Linn Edsall of Harvard* (Boston: Harvard Medical Alumni Association, 1970), p. 206.

4. In 1930 a medical service was also assigned to Boston University and by 1940 each school had been allotted two medical services. Byrne, *History of Boston City Hospital,* pp. 93, 96.

5. Board of Trustees and Staff of the Boston City Hospital, *Report of the Joint Committee to Prepare Plans for the Future Development of the Hospital* (privately printed, 1916), p. 28; Editorial, "Continuous Service at the Boston City Hospital," *Boston Medical and Surgical Journal* 171 (October 8, 1914):573; Byrne, *History of Boston City Hospital,* pp. 42–43; Vogel, *Invention of the Modern Hospital,* pp. 81–82; "Dedication: Cheever Amphitheatre, Sears Surgical Laboratory, Boston City Hospital, May 25, 1966," program, available at Boston City Hospital Library.

6. James M. Curley, "The Thorndike Memorial," *Boston Medical and Surgical Journal* 189 (November 15, 1923):734; Townsend W. Thorndike, "Address" (at Thorndike Memorial Exercises), ibid., p. 740; Obituary, "Francis Weld Peabody," *Journal of Clinical Investigation* 5 (December 1927):1–3.

7. C. Sidney Burwell to Alan M. Chesney, December 15, 1941, Alan M. Chesney Archives, Johns Hopkins University School of Medicine, Baltimore, Md.

8. Byrne, *History of Boston City Hospital,* pp. 160, 186–187.

9. Ibid., pp. 162, 190–191; John J. Byrne to Harry F. Dowling, April 1, 1981.

10. Minutes of the Surgical Staff, Boston City Hospital, December 18, 1928, available at Boston City Hospital Library. Around this time the surgical staff also prevented the appointment of the chairman of the Tufts Department of Surgery, Stuart Welsh, as an attending surgeon. John J. Byrne to Harry F. Dowling, April 1, 1981.

11. Byrne, *History of Boston City Hospital,* pp. 191–194.

12. Some implied that those in favor of introducing university services may have instigated the association's action. John J. Byrne to Harry F. Dowling, and accompanying material, January 19, 1981.

13. John J. Byrne to Harry F. Dowling, and accompanying material, January 19, 1981; Byrne, *History of Boston City Hospital,* pp. 165–166, 169, 178–183, 285; Castle, "Harvard at the Boston City Hospital," pp. 11, 17.

14. George Packer Berry to Richard J. Condon, March 4, 1954, Berry Papers, Francis A. Countway Library of Medicine, Boston, Mass.; *Boston Herald American,* May 10, 1955; William B. Castle, "Harvard at the Boston City Hospital," reprinted from Aesculapian Club of Boston, *1964 Bulletin,* n.p., p. 11; Byrne, *History of Boston City Hospital,* pp. 195–196. By the academic year of 1964–65 Harvard Medical School was supplying $1,695,000 for teaching, research, and patient care on all the Harvard services at Boston City Hospital, compared with approximately $500,000 from the hospital's budget. "Remarks by Dean George P. Berry at the Reopening of the Thorndike Memorial Laboratory," in *Training of the Physician: Papers*

*Presented as Part of the Celebration of the Centennial of the Boston City Hospital and the Fortieth Anniversary of the Thorndike Memorial Laboratory* (Boston, 1964), p. 120.

15. The quotation is from Harvey Cushing, *The Life of Sir William Osler*, 2 vols. (Oxford: Clarendon Press, 1925), 1:313n. See also George S. Mirick, "Medical Department of the Baltimore City Hospitals," *Maryland State Medical Journal* 4 (December 1955):759; Douglas Carroll, "History of the Baltimore City Hospitals," ibid. 15 (July 1966):117; Alan M. Chesney, "Memorandum re Relations of the Johns Hopkins University School of Medicine to the City Asylum at Bay View," January 27, 1961, Alan M. Chesney Archives, Johns Hopkins University, Baltimore, Md.; Abou Pollack, "A Brief History of the Pathology Laboratory of the Baltimore City Hospitals," *Maryland State Medical Journal* 4 (December 1955):781.

16. Alan M. Chesney, *The Johns Hopkins Hospital and the Johns Hopkins University School of Medicine: A Chronicle*, 3 vols. (Baltimore: Johns Hopkins Press, 1943, 1958, 1963), vol. 1, *1867–1893*, pp. 107–108; Peter D. Olch, "William S. Halsted and Local Anesthesia," *Anesthesiology* 42 (April 1975):483–486.

17. Simon Flexner and James Thomas Flexner, *William Henry Welch and the Heroic Age of American Medicine* (New York: Viking Press, 1941), pp. 128–134; Cushing, *Osler*, 1:297; Olch, "William S. Halsted and Local Anesthesia," pp. 484–485; Chesney, *Johns Hopkins Hospital*, vol. 3, *1905–1914*, p. 118; Thomas B. Turner, *Heritage of Excellence: The Johns Hopkins Medical Institutions, 1914–1917* (Baltimore: Johns Hopkins University Press, 1974), p. 304.

18. Warfield T. Longcope, "Thomas Richmond Boggs, 1875–1938," *Transactions of the Association of American Physicians* 54 (1939):9; Chesney, *Johns Hopkins Hospital*, vol. 2, *1893–1905*, pp. 415–416. The quotation of Osler is cited in Carroll, "History of the Baltimore City Hospitals," August 1966, p. 69.

19. Chesney, *Johns Hopkins Hospital*, 3:119–120; "Arthur Marriott Shipley, 1878–1955," *The Record* (Baltimore) 3 (September 1960):4.

20. Chesney, *Johns Hopkins Hospital*, 3:119; *Who Was Who in America* (Chicago: A.N. Marquis, 1960), vol. 3, *1951–1960*, p. 931; Pollack, "History of the Pathology Laboratory at Baltimore City Hospitals," p. 781; *University of Virginia Record*, cat. no. 18 (March 15, 1932):19; Duke University, *The First Twenty Years: A History of the Duke University Schools of Medicine, Nursing and Health Services, and Duke Hospital, 1930 to 1950*, Bulletin of Duke University 24 (May 1952):104; Turner, *Heritage of Excellence*, pp. 519–520.

21. City of Baltimore, Department of Charities and Correction, Supervisors of City Charities, *Twenty-ninth Annual Report* (Baltimore, 1928), p. 32; Alan M. Chesney, "Excerpts from Dean's Report, 1932, 33," p. 2, Alan M. Chesney Archives; Board of Supervisors of the Baltimore City Hospitals, Minutes, February 9, 1934, Maryland Historical Society, Baltimore, Md.

22. The quotation is from Karl A. Meyer, "Frederick Tice, 1871–1953," *Proceedings of the Institute of Medicine of Chicago* 20 (May 15, 1954):126. See also Carroll, "History of the Baltimore City Hospitals," October 1966, p. 90.

23. New York University, *Bulletin of the School of Medicine and Post-Graduate School, 1963–64*, p. 45; New Jersey Medical School, *Catalog of College of Medicine and Dentistry of New Jersey, 1976–78*, p. 51; *Who's Who in America*, 40th ed. (Chicago: Marquis Who's Who, 1978–79), pp. 534, 590, 1842, 2821.

24. James A. Gannon, "Recollections of Old Gallinger Hospital," *Medical Annals of the District of Columbia* 20 (July 1951):391, 393.

25. District of Columbia, Board of Public Welfare, *Annual Report, 1927* (Washington, D.C.: U.S. Government Printing Office, 1927), p. 54; ibid., *1929*, pp. 66, 67, 73; Council on Medical Education and Hospitals of the American Medical Association, "Hospital Service in the United States, 1928," *Journal of the American Medical Association* 90 (March 24, 1928):930; Council on Medical Education and Hospitals of the American Medical Association, "Hospital Service in the United States, 1929," ibid. 92 (March 30, 1929):1066.

26. *Medical Annals of the District of Columbia* 1 (February 1932):48.

27. U.S. Congress, House, *Congressional Record, 78*th Cong., 1st sess., October 19, 1943, pp. 8542–8543, remarks of Mr. Hancock, New York; District of Columbia, *Report of the Government, 1938* (Washington, D.C.: U.S. Government Printing Office, 1938), p. 165.

28. They were Albert J. Sullivan, J. Ross Veal, John Parks, Lewis K. Sweet, and Charles P. Cake, respectively.

29. The staff for 1932 is listed in *Medical Annals of the District of Columbia* 1 (February 1932):48. See also *Washington Post,* February 19, 1939.

30. District of Columbia, *Report of the Government, 1939* (Washington, D.C.: U.S. Government Printing Office, 1939), p. 201; *Washington Star,* February 17, 1939. It will be recalled that medical schools in Philadelphia and New York had had similar difficulties in the past when city authorities appointed chiefs of services to supersede the attending staff in directing the care of patients (see Chapter 1). The situations differed, however, in that those doctors were political hacks not capable of exercising the authority given them, while the chief medical officers at Gallinger Municipal Hospital were academically trained and dedicated doctors.

31. District of Columbia, *Report of the Government, 1939,* pp. 201–202; ibid., *1940,* p. 196; *Washington Post,* February 19, 1939; *Washington Star,* February 26, 1939, March 12, 1939, April 29, 1939; Albert J. Sullivan, "Recent Changes in the Medical Service at Gallinger," *Health in the Capital City* 1 (February 25, 1939):1–2.

32. Walter C. Hess, "The History of the Georgetown University School of Medicine, 1930–1964," *Georgetown Medical Bulletin* 18 (August 1964):57–58.

33. Joseph A. Capps, "Efficiency Tests Applied to the Attending and House Physicians of the Cook County Hospital, Chicago," *Boston Medical and Surgical Journal.* 177 (August 30, 1917):286–287; Editorial, "Efficiency at the County Hospital," *Bulletin of the Chicago Medical Society* 14 (November 7, 1914):16.

34. "Plan of Re-organization of Cook County Hospital," *Bulletin of the Chicago Medical Society* 33 (January 10, 1931):16–18; William F. Petersen, "The Problem of the Cook County Hospital Staff Reorganization," abstract of address delivered at the Chicago Medical Women's Club, March 11, 1931, n.p., available at Health Sciences Library, Abraham Lincoln School of Medicine, University of Illinois, Chicago; Charles A. Elliott, "The Spirit of the Clinic," *Bulletin of the Northwestern University Medical School* 37 (July 12, 1937):4.

35. The gap between the supervision required by the rapidly advancing science of medicine and the amount provided by the overextended attending staff, never openly admitted by hospital authorities, was unwittingly revealed in 1949 by two attending surgeons reporting on the management of

obstruction of the small intestine. They explained that over three hundred patients with this serious and potentially fatal complication came to Cook County Hospital each year and that "members of the House Staff are responsible for almost all the care of these patients, guiding pre- and postoperative management and performing the major part of the surgery, assisted by the Cook County Hospital Night Surgeon, who is usually an associate or younger staff member." Samuel J. Fogelson and Chester Moen, "Management of Mechanical Obstruction of the Small Intestine at Cook County Hospital," *Quarterly Bulletin of the Northwestern University Medical School* 23 (Fall 1949):287.

36. Malcolm T. MacEachern, "Summary Report of Findings and Recommendations from the Survey of the Cook County Hospital and Cook County School of Nursing" (mimeographed, Chicago, 1932), pp. 11–17.

37. "Report of Committee on Survey of Cook County Hospital," *Chicago Hospital Council Bulletin* 1 (September 1938):9–24; "Summary of Medical Activities in Cook County Hospital: A Report by the Board of Commissioners of Cook County," ibid., October 1938, p. 11.

38. Raymond W. McNealy, "The Influence of Cook County Hospital on Medical Education in the United States," *Quarterly Bulletin of the Northwestern University Medical School* 31 (Summer 1957):169–173; Charles B. Johnson, *Growth of Cook County* (Chicago: Board of Commissioners of Cook County, Ill., 1960), pp. 170–171, 185, 187–191; "Cook County Hospital," *Resident Physician* 1 (September 1955):52–63; Marvin J. Colbert to Harry F. Dowling, April 25, 1958.

39. S. Howard Armstrong, "The Cook County Picture and the National Picture on Internships: Report to Dr. Karl A. Meyer and the Civil Services Commission of Cook County" (mimeographed, n.d.); Citizens Committee on Cook County Government, "Interim Report: Health and Hospital Services" (mimeographed, Chicago, 1967), p. 17. While standards were high in the medical schools of several foreign countries, most foreign-trained house officers applying for positions in the United States after the Second World War were graduates of schools that were short of funds, equipment, and adequate clinical facilities and where the teacher-student ratio was low as compared to schools in the United States and Canada. Thus, leading hospitals, public and private, usually confined their appointments of house officers to graduates of schools in these two countries plus the few who applied from other countries with high standards of medical education.

40. "Curriculum Vitae: Edmund F. Foley," files of the Department of Medicine, Abraham Lincoln School of Medicine, University of Illinois, Chicago.

41. Citizens Committee on Cook County Government, "Interim Report," p. 18; Abraham Flexner, *Medical Education in the United States and Canada: A Report to the Carnegie Foundation for the Advancement of Teaching* (New York: Carnegie Foundation for the Advancement of Teaching, 1910), pp. 275–278; Alexander M. Schmidt to Harry F. Dowling, June 6, 1981.

42. "Here and There in the Medical Field," *Chicago Medical Society Bulletin* 54 (March 8, 1952):645; Editorial, "The Future of Cook County Hospital," *Chicago Medicine* 67 (May 30, 1964):437; Editorial, "Progress at Cook County Hospital," ibid. 66 (June 15, 1963):532.

43. Maxwell Finland, "The Harvard Medical Unit of the Boston City Hospital," *New England Journal of Medicine* 271 (November 19, 1964):1096–1100.

44. Conversation with Anne G. Hargreaves, Boston, Mass., December 4, 1980.

45. Alan M. Chesney, "Memorandum re Relations of the Johns Hopkins University School of Medicine to the Baltimore City Hospitals," January 31, 1961, p. 4, Alan M. Chesney Archives; Alan M. Chesney, "Chronology of Relations of Johns Hopkins University School of Medicine to 'Bay View,'" January 31, 1961, pp. 6–7, Alan M. Chesney Archives; Mirick, "Medical Department of the Baltimore City Hospitals," p. 761; conversation with A. McGehee Harvey, Baltimore, Md., March 6, 1980.

46. Conversation with Theodore Woodward, Atlanta, Ga., October 6, 1978; Parker J. McMillin, "Baltimore City Hospitals," *Maryland State Medical Journal* 4 (December 1955):751–752; A.M. Shipley and O.C. Brantigan, "The Surgical Service of Baltimore City Hospitals," ibid., p. 768.

47. *Baltimore Sun,* August 24, 1961, August 30, 1961; *Baltimore Evening Sun,* August 30, 1961; Baltimore Department of Public Welfare, *Twenty-seventh Annual Report* (Baltimore, 1961), pp. 27–28.

48. The quotation is from Press Conference, December 21, 1961, cited in *Baltimore Health News* 39 (February 1962): 13. See also "Minutes of the 1st Meeting of the Baltimore City Advisory Committee on Medical Care," August 22, 1947, Alan M. Chesney Archives; "Memorandum of Agreement" between Baltimore Board of Public Welfare and Johns Hopkins University, February 13, 1961, Alan M. Chesney Archives; *Baltimore Sun,* August 24, 1961, August 30, 1961; *Baltimore Evening Sun,* August 30, 1961.

49. Rudolph Matas, "Dr. Charles C. Bass, Dean: An Appreciation," *New Orleans Medical and Surgical Journal* 92 (April 1940):549–550; "Double Bed Charity," *Time,* November 29, 1937, p. 53.

50. "1916–1926: Decade of Great Progress," *Hospital Management* 21 (February 1926):28; John E. Deitrick and Robert C. Berson, *Medical Schools in the United States at Mid-Century* (New York: McGraw-Hill, 1953), p. 145.

51. The general municipal hospitals affiliated with medical schools were Harlem with Columbia, Lincoln with Einstein, and Elmhurst City Hospital Center with Mount Sinai Hospital, which was in the process of establishing a medical school. Bellevue Hospital Center remained affiliated with New York University and Kings County Hospital Center with the Downstate Medical Center of the State University of New York. General hospitals that affiliated with voluntary hospitals were Gouverneur, Morrisania, Greenpoint, Cumberland, Coney Island, Queens General, and Fordham. Robb K. Burlage, *New York City's Municipal Hospitals: A Policy Review* (Washington, D.C.: Institute for Policy Studies, 1967), p. 121.

52. Burlage, *New York City's Municipal Hospitals,* pp. 33–34, 78, 132, 133; the quotations are on pp. 33 and 132.

53. Ibid., pp. 210, 214–215, 216a, 218–220, 222a.

54. Ibid., pp. 384–391, 427–430.

55. John C. Munro, "Surgery of the Vascular System: I. Ligation of the Ductus Arteriosus," *Annals of Surgery* 46 (September 1907):335–338. Since diagnosis during life required techniques not available at that time, the actual operation was not performed successfully until thirty years later (by Robert Gross at the Children's Hospital of Boston). Stephen L. Johnson, *The History of Cardiac Surgery, 1896–1955* (Baltimore: Johns Hopkins Press, 1970), pp. 73–74.

56. Dowling, *Fighting Infection,* p. 5; George Derby, "Carbolic Acid in Surgery," *Boston Medical and Surgical Journal* 77 (October 24, 1867):

271–273; J. Collins Warren, *To Work in the Vineyard of Surgery,* ed. Edward D. Churchill (Cambridge, Mass.: Harvard University Press, 1958), p. 138n; Cheever et al., *History of Boston City Hospital,* p. 282. The journal was the *Boston City Hospital Medical and Surgical Reports.*

57. Dowling, *Fighting Infection,* pp. 48–49, 113, 121, 141–142, 157, 188.

58. Jerome O. Klein and Edward H. Kass, eds., "The Art of Clinical Investigation: Current Reviews of the Contributions of Maxwell Finland," *Journal of Infectious Diseases* 125, supp. (March 1972).

59. George W. Corner, *A History of the Rockefeller Institute, 1901–1953* (New York: Rockefeller Institute Press, 1964), pp. 1–55 passim; "Medical News," *Journal of the American Medical Association* 49 (July 6, 1907):51; Lewellys F. Barker, "Theodore Caldwell Janeway," *Science* 47 (March 22, 1918):275; Morris Bishop, *A History of Cornell* (Ithaca, N.Y.: Cornell University Press, 1962), p. 386; Cornell University, *Announcement of the Medical College for 1933–34* (Ithaca, N.Y., 1933), p. 49; Graham Lusk, "Clinical Calorimetry: A Respiration Calorimeter for the Study of Disease," *Archives of Internal Medicine* 15 (May 1915):793–804; Frank C. Gephart and Eugene F. DuBois, "Clinical Calorimetry: The Organization of a Small Metabolism Ward," ibid., pp. 829–867; John Duffy, *A History of Public Health in New York City, 1866–1966* (New York: Russell Sage Foundation, 1974), p. 262; Dowling, *Fighting Infection,* pp. 43–44.

60. Curley, "Thorndike Memorial," p. 736; Editorial, "The Thorndike Memorial Laboratory," *New England Medical and Surgical Journal* 189 (November 15, 1923):785.

61. W.B. Castle, "Thorndike Memorial Laboratory," n.d., George Packer Berry Papers, Francis A. Countway Library of Medicine, Boston; Byrne, *History of Boston City Hospital,* pp. 70–71; Maxwell Finland, "The Harvard Medical Unit," in *Training of the Physician,* pp. 131–134; George R. Robb, comments, in Boston City Hospital, *The Harvard Medical Unit: An Autobiography, 1915–1964* (Boston, n.d.), p. 11; Klein and Kass, "The Art of Clinical Investigation."

62. The quotation is from Editorial, "Thorndike Memorial Laboratory," p. 785. See also Byrne, *History of Boston City Hospital,* pp. 49, 51.

63. Between July 1, 1965, and June 30, 1967, funds available for research at Boston City Hospital were, in thousands of dollars:

| School | Grant funds (by applicant) | | |
|--------|-------------------|--------------------|-------|
| | *Medical schools* | *Boston City Hospital* | *Total* |
| Harvard | $3,995 | $743 | $4,738 |
| Boston University | 267 | 541 | 801 |
| Tufts | 444 | 432 | 876 |

*Source:* City of Boston, Department of Health and Hospitals, *Annual Report,* 1966 (Boston, 1966), p. 15.

64. Thomas Neville Bonner, *Medicine in Chicago, 1850–1950: A Chapter in the Social and Scientific Development of a City* (Madison, Wis.: American History Research Center, 1957), pp. 103–104; Karl A. Meyer, "Historical Background of Cook County Hospital," *Quarterly Bulletin of Northwestern University Medical School* 23 (Fall 1949):274; Ethel H. Davis, *The Hektoen*

*Institute for Medical Research* (Chicago: Kuttner and Kuttner, n.d.), p. 5; "The Hektoen Institute," *Chicago Medicine* 67 (December 12, 1964):1040–1041.

65. New York City Department of Hospitals, *Research: An Experiment in Municipal Organization, Goldwater Memorial Hospital* (New York, 1945); David P. Earle to Harry F. Dowling, and accompanying material, June 27, 1980.

66. N.W. Shock, "The United States Public Health Service-Baltimore City Hospitals Research Section on Gerontology," *Baltimore Health News* 26 (March 1949):106–112; U.S. Department of Health, Education and Welfare, Public Health Service, National Institutes of Health, *To Understand the Aging Process: The Baltimore Longitudinal Study of the National Institute on Aging,* DHEW Publication no. (NIH) 78–134 (Washington, D.C.: U.S. Government Printing Office, n.d.); U.S. Department of Health, Education and Welfare, *National Institute on Aging,* DHEW Publication No. 78–1129 (Washington, n.d.); "Ten Million Dollar Research Building," *The Record* (Baltimore) 4 (July-August 1962):2; "Dedication of the Gerontology Research Center at Baltimore City Hospitals," ibid. 10 (June 1968):1, 3–4.

67. Stephen L. Johnson, *The History of Cardiac Surgery, 1896–1955* (Baltimore: Johns Hopkins Press, 1970), pp. 129–136; Dickinson W. Richards, "Right Heart Catheterization," Nobel Lecture, 1956, in Dickinson W. Richards, *Medical Priesthoods and Other Essays* (privately published, 1970), pp. 85–98.

68. Joseph A. Capps and George H. Coleman, *An Experimental and Clinical Study of Pain in the Pleura, Pericardium and Peritoneum* (New York: Macmillan, 1932).

69. Karl A. Meyer, "The History of the Cook County Hospital Blood Bank," *Quarterly Bulletin of the Northwestern University Medical School* 23 (Fall 1949):318–320; Dowling, *Fighting Infection,* pp. 264–265.

70. Mirick, "Medical Department of the Baltimore City Hospitals," pp. 758–762.

71. Conversation with Sol Katz, Baltimore, Md., June 1, 1981; conversation with Max M. Montgomery, Chicago, Ill., October 31, 1979.

72. Jan de Hartog, *The Hospital* (New York: Atheneum, 1964), pp. 120–121, 163–164.

73. Ibid., p. 146; Francis W. Peabody, "The Care of the Patient," *Journal of the American Medical Association* 88 (March 19, 1927):877–882; the quotation is on p. 882.

74. Emil Bogen, "Intern Year Appreciated," *Hospital Management* 18 (September 1924):68; Page Cooper, *The Bellevue Story* (New York: Thomas Y. Crowell, 1948), p. 4.

75. Boston City Hospital, *Harvard Medical Unit.*

76. The figures are from Minutes of the 187th Meeting of the Medical Advisory Board, Baltimore City Hospitals, May 1, 1952, Alan M. Chesney Archives. The quotation was reported in a conversation with William Jordan, Washington, D.C., May 10, 1979.

77. Aub and Hapgood, *Pioneer in Modern Medicine,* p. 160; Eugene F. DuBois and Paul Reznikoff, "The Clinical Clerkship in Medicine: Details of a Method of Instruction," *Journal of the American Medical Association* 87 (August 21, 1926):542–545.

78. Conversation with Max M. Montgomery, October 31, 1979.

79. The four professors were Arthur R. Colwell, Northwestern; Harry F. Dowling, Illinois; Peter J. Talso, Loyola; and Hyman J. Zimmerman, Chicago Medical School.

80. Report of the Committee to Investigate the Care of Private Patients at the Baltimore City Hospitals, June 20, 1960, Mayor's Administrative Files, Record Group 9, Baltimore City Archives.

## 7. Medical Care: Running to Keep Up

Epigraphs: *Chicago Daily News,* August 2, 1941; Lewis Thomas, "Public Teaching Hospitals Face Life-Death Issues in Changing Times," *Modern Hospital* 109 (October 1967):95.

1. Robert D. Grove and Alice M. Hetzel, *Vital Statistics in the United States, 1940–1960,* Public Health Service Publication no. 1677 (Washington, D.C.: U.S. Government Printing Office, 1968), pp. 80–84, 95, 309; Harry F. Dowling, *Fighting Infection: Conquests of the Twentieth Century* (Cambridge, Mass.: Harvard University Press, 1977), p. 230.

2. Jan de Hartog, *The Hospital* (New York: Atheneum, 1964).

3. "Report of Committee on Survey of Cook County Hospital," *Chicago Hospital Council Bulletin* 1 (September 1938):16; De Hartog, *The Hospital,* pp. 27–34.

4. William A. Nolen, "Happy Days at Bellevue," *Esquire,* November 1969, p. 171.

5. The ratio of nurses to beds was a designation commonly employed in the past, but its rigid application resulted in short staffing, since one nurse is needed per eight-hour shift, not one nurse per twenty-four-hour shift, to achieve full staffing. Because one nurse works five shifts per week, and there are twenty-one shifts in a week, 4.2 nurses on the payroll per week are needed to achieve the services of one nurse at all times. Thus, the nurse-to-bed ratio could give the impression that over four times as many nurses were available as were actually on duty at one time. Conversation with Mary Mullane, Silver Spring, Md., June 18, 1981.

6. Malcolm T. MacEachern, "Summary Report of Findings and Recommendations from the Survey of the Cook County Hospital and Cook County School of Nursing" (mimeographed, Chicago, 1932), pp. 21–22, 131; S.H. Rosenblum, "History of the Cook County Hospital" (mimeographed, Chicago, 1953), p. 8; Elizabeth Jane Davis, "History, Development and Organization: Cook County Hospital," *Chicago Hospital Council Bulletin* 8 (September 1945):2.

7. Richard Harrison Shryock, "Nursing Emerges As a Profession: The American Experience," in *Sickness and Health in America: Readings in the History of Medicine and Public Health,* ed. Judith Walzer Leavitt and Ronald L. Numbers (Madison, Wis.: University of Wisconsin Press, 1978), p. 208; MacEachern, "Summary Report of Findings," pp. 21–23; "Report of Committee on Survey of Cook County Hospital" (mimeographed, Chicago, 1937), pp. 13–15.

8. James A. Hamilton and Associates, "Study of Nursing of the Cook County School of Nursing in the Cook County Hospital, Chicago, Illinois" (mimeographed, Chicago, 1950), pp. I–1, I–2, III–11, III–12, III–19.

9. Ibid., p. II–52; Citizens' Committee to Study Cook County Hospital, "Final Report" (mimeographed, Chicago, 1964), Tab B, pp. 2–3. The city-county hospitals included in the average in 1949 are Charity of New Orleans, Cleveland City, Los Angeles County, Anker (St. Paul), Cincinnati General, Minneapolis General, Alameda County (Oakland), and Detroit Receiving. The state-owned hospitals are located at the universities of Minnesota, Nebraska, Iowa, Illinois, Michigan, Indiana, and Wisconsin.

10. The nursing ratios are from Board of Trustees and Staff of the Boston City Hospital, *Report of the Joint Committee to Prepare Plans for the Future Development of the Hospital* (privately printed, 1916), p. 19. The quotation is from Baltimore Board of Supervisors of City Charities, *Annual Report for the Fiscal Year Ending December 31, 1926* (Baltimore: Department of Charities and Correction, 1926), Exhibit A, pp. 24–25.

11. *Baltimore Sun,* September 7, 1933. The quotation is from Commission on Governmental Efficiency and Economy, Inc., Baltimore, "Baltimore City Hospitals: A Study of the Business Aspects of Operation and Management" (mimeographed, Baltimore, February 1949), p. 14. See also LeRoy E. Hoffberger to Thomas J. D'Alesandro, III, September 2, 1970, Mayor's Administrative Files, Record Group 9, Baltimore City Archives.

12. The National League of Nursing was founded in 1894 and the American Nurses' Association in 1896. Shryock, "Nursing Emerges As a Profession," p. 207.

13. Ibid., pp. 205–206, 208, 209, 210.

14. Isabel Maitland Stewart, *The Education of Nurses: Historical Foundations and Modern Trends* (New York: Macmillan, 1943), pp. 225–227, 276–277, 290–291; Philip A. Kalisch and Beatrice J. Kalisch, *The Advance of American Nursing* (Boston: Little, Brown, 1978), p. 652.

15. Daniel L. Seckinger "A Brief History of District of Columbia General Hospital" (mimeographed, Washington, D.C., n.d.), pp. 7–8, 11–12.

16. *Washington Post,* February 17, 1936.

17. The quotations are from Dorothy Jane Youtz, *The Capital City School of Nursing, Formerly the Washington Training School for Nurses, 1877–1972,* ed. Irene B. Page and Mary M. Goodreau (Washington, D.C.: Capital City School of Nursing Alumni Association, 1975), p. 100; and U.S. Public Health Service, *Report of a Survey of the Health Department and Other Health Agencies in the District of Columbia* (Washington, D.C.: U.S. Government Printing Office, 1939), p. 90. See also U.S. Congress, Senate, Committee on the District of Columbia, *Conditions at Gallinger Municipal Hospital,* 78th Cong., 1st sess., September 27-October 1, 1943 (Washington, D.C.: U.S. Government Printing Office, 1943): Isadore and Zachary Rosenfeld and John Steinle and Associates, "Survey of the Municipal Hospital Facilities of the District of Columbia: Summary Report," *Medical Annals of the District of Columbia* 27 (February 1958):91–99; *Washington Post,* April 17, 1966.

18. The quotation is from Rosenfeld et al., "Survey of Municipal Hospital Facilities," p. 92. See also Daniel Seckinger, *Highlights for 1951: Gallinger Municipal Hospital* (Washington, D.C.: District of Columbia Health Department, n.d.), n.p.

19. Youtz, *Capital City School of Nursing,* p. 216.

20. Hamilton and Associates, "Study of Nursing in Cook County Hospital," pp. II–26, II–28, II–29.

21. Ibid., pp. I–5, II–36–II–39; "Medical News: New Nursing School for County Hospital," *Journal of the American Medical Association* 93 (August 24, 1929):618; James E. Hague and Earl V. Fischer, "Cook County: The Top of the Team," *Hospitals* (February 1974):66; "Diploma Programs in Nursing Accredited by the NLN, Sept. 16, 1979-Sept. 15, 1980," *Nursing Outlook* 27 (August 1979):549–553.

22. Conversation with Anne G. Hargreaves, Boston, Mass., December 4, 1980.

23. "Shortage of Freshman Students at Boston Nursing Schools Reported," *Modern Hospital* 109 (August 1967):30.

24. Conversation with Edmund G. Beacham, Baltimore, Md., April 1, 1981.

25. See, for instance, the speech of Congresswoman Frances P. Bolton in 1946, in which she gave an impassioned plea for a more responsible attitude and larger appropriations from Congress for Gallinger Hospital. *Congressional Record,* Appendix, June 26, 1946, pp. A3743–3745.

26. "Interim Classification of Schools of Nursing Offering Basic Programs (1949)," *National Journal of Nursing* 49 (November 1949):34–46.

27. "Report of Committee on Survey of Cook County Hospital," pp. 17–20.

28. Citizens Committee on Cook County Governance, "Interim Report: Health and Hospital Services" (mimeographed, Chicago, 1967), p. 15; "Cook County Hospital under Fire from Nurses and Citizens Group," *Modern Hospital* 110 (January 1968):42, "Cook County Hospital Faces Crisis over Accreditation and Hiring of Administrator," ibid. 102 (May 1964):176–177; "To Eliminate 40-Bed Wards at Cook County Hospital," *Hospitals* 28 (May 1954):151.

29. Rolf M. Gunnar to Sheldon S. Waldstein, April 14, 1966, copy in author's possession; George D. Benton to Rolf M. Gunnar, September 18, 1967, copy in author's possession; "Report of the Ad Hoc Committee of the Academic Assembly To Be Submitted to the Executive Committee of the Cook County Hospital for Transmission to the President of the Board of Commissioners of Cook County" (mimeographed, Chicago, January 4, 1968), p. 3, copy in author's possession.

30. U.S. Congress, Senate, Committee on the District of Columbia, Subcommittee on Public Health, Education, Welfare and Safety, *District of Columbia Hospital and Health Facility Construction and Modernization.* 90th Cong., 1st sess., August 21, 1967, p. 91.

31. Conversation with Maxwell Finland, Boston, Mass., December 3, 1980.

32. Frances Burns, "Why Boston Can't Afford to Lose the City Hospital," series of eleven articles in *Boston Globe,* November 16–26, 1958, reprinted by city of Boston, December 1958, pp.9, 11–12, 40.

33. Frank Smith, "Misery Harbor: What Goes on behind County Hospital Doors," *Chicago Times,* series of articles, February-March 1936; Fitzhugh Mullan, "The Butcher Shop," *Hospital Physician* 12 (October 1976):16–19; ibid. (November 1976):28–32; *Washington Post,* April 13, 1946.

34. "Can't Meet Commission's Tommyrot, Dr. Meyer Says," *Modern Hospital* 100 (April 1963):170; "William M. McCoy Named to $30,000 Post As Head of Troubled Cook County Hospital," ibid. 104 (April 1965):29.

35. T.J.S. Waxter to Thomas D'Alesandro, Jr., March 10, 1949, in Mayor's Administrative Files, Group 9, Baltimore City Archives; *Baltimore Sun,* July 17, 1966.

36. U.S. Congress, Senate, *Congressional Record,* November 28, 1944, remarks of Mr. Mason, p. 8477.

37. De Hartog, *The Hospital,* p. 35.

38. "Report to the Staff (of the Children's Division, Cook County Hospital)" (mimeographed, Chicago, 1965), p. 3; Citizens Committee on Cook County Governance, "Interim Report," p. 15. See also "Hospital News," *Journal of the American Medical Association* 106 (March 14, 1936):928; Kenneth B. Babcock to Fred A. Hertwig [December 1962], Cook County Hospital Archives.

39. *Report of the Government of the District of Columbia for the Year Ended June 30, 1943* (Washington, D.C.: Division of Printing, D.C. Government, n.d.), pp. 157–158; ibid., *June 30, 1949* (Washington, D.C.: U.S. Government Printing Office, 1950), p. 191.

40. The information on dumping is from Minutes of the Board of Supervisors of City Charities, Baltimore, Maryland, April 4, 1924, Maryland Historical Society, Baltimore; *Baltimore Evening Sun*, November 3, 1959; conversation with Philip Zieve, Baltimore, Md., February 19, 1980; conversation with Edmund G. Beacham, Baltimore, Md., November 6, 1980. The statistics are from Baltimore Department of Public Welfare, *Third Annual Report* (Baltimore: King Brothers, 1937), p. 137; Baltimore Hospital Survey Committee, "General Hospitals for the Baltimore Area" (mimeographed, Baltimore, 1958), p. 38; *Report of the Government of the District of Columbia for the Year Ended June 30, 1933* (Washington, D.C.: U.S. Government Printing Office, 1934), p. 103; *Washington Post*, May 10, 1961; Robb K. Burlage, *New York City's Municipal Hospitals: A Policy Review* (Washington, D.C.: Institute for Policy Studies, 1967), p. 28.

41. S.S. Goldwater, "The United States Hospital Field," *National Hospital Record* 9 (April 1906):11; *Report of the Government of the District of Columbia for the Year Ended June 30, 1938* (Washington, D.C.: U.S. Government Printing Office, 1938), p. 169.

42. "Should Municipal and County Hospitals Be Wholly Free?" *Modern Hospital* 25 (October 1925):327–328.

43. C.A. Alexander, "The Evolution of Public Medical Care in Baltimore City," *Maryland State Medical Journal* 16 (June 1967):61; U.S. Department of Commerce, Bureau of the Census, *Historical Statistics of the United States: Colonial Times to 1970*, 2 vols. (Washington, D.C.: U.S. Government Printing Office, 1975), 1:82.

44. The quotation is from Baltimore Department of Public Welfare, *Fourth Annual Report* (Baltimore: King Brothers, 1938), pp. 160–161. See also Baltimore Department of Welfare, *Seventeenth Annual Report* (Baltimore, 1951), p. 33; Parker J. McMillin, "Baltimore City Hospitals," *Maryland State Medical Journal* 4 (December 1955):753–758; Health Insurance Institute, *1968 Source Book of Health Insurance Data* (New York, 1968), p. 16.

45. *Baltimore Sun*, June 30, 1960; Conrad Acton, "Baltimore City Medical Society," *Maryland State Medical Journal* 7 (December 1958):706–708; ibid. 9 (August 1960):457–461; Milton Golin, "How Baltimore 'Rescued' Its City Hospitals," *American Medical News*, October 26, 1979, pp.1–3.

46. Golin, "How Baltimore 'Rescued' Its City Hospitals."

47. *Report of the Government of D.C., Year Ended June 30, 1943* (Washington D.C.: U.S. Government Printing Office, 1944), pp. 156, 159; ibid., *Year Ended June 30, 1947* (Washington D.C.: U.S. Government Printing Office, 1948), p. 186; *Washington Star*, February 1, 1962; U.S. Congress, Senate, *Congressional Record*, October 14, 1969, pp. S12534–S12535.

48. "Should Municipal Hospitals Be Wholly Free?"

49. See, for example, *Baltimore Sun*, June 28, 1931.

50. *Baltimore Evening Sun*, April 17, 1935; Minutes of the Board of Supervisors of City Charities, Baltimore, Maryland, December 9, 1932; Baltimore Department of Public Welfare, *First Annual Report* (Baltimore: Daily Record Press, 1935), p. 44; Baltimore Council of Social Agencies, "Chronic Hospital Care: A Study of Need and Resources for Chronic Hospital Care in Baltimore" (mimeographed, Baltimore, 1940), p. 37, copy in

Legislative Reference Files, Record Group 29, Baltimore City Archives; "An Assistant Director of Welfare for Medical Care," *Baltimore Health News* 21 (June 1944):44; "Interim Report on Medical Care for Baltimore City," ibid. 24 (January-February-March 1947):102. The quotation is from W. Ross Cameron, "The First Year's Work of the Assistant Director of Welfare for Medical Care," ibid. 23 (July 1946):50.

51. Brochure of Baltimore City Hospitals (mimeographed, Baltimore, 1968), copy in Mayor's Administrative Files, Record Group 9, Baltimore City Archives; Frederick G. Hubbard, "City Hospitals," in Mayor's Annual Reports, 1963–67, p. 73, Baltimore City Archives; Douglas Carroll, "History of the Baltimore City Hospitals," *Maryland State Medical Journal* 15 (October 1966):95; "The Evolution of a Hospital," *The Record* (Baltimore) 7 (May-June 1965):3–4; Ernst P. Boas, "Where Sick Poor Are Hospitalized," *Hospital Management* 20 (July 1925):35–36.

52. Hubbard, "City Hospitals," p. 77; U.S. Congress, Senate, Committee on Labor and Public Welfare, Subcommittee on Problems of the Aged and Aging, *The Condition of American Nursing Homes,* 80th Cong., 2d sess. (Committee Print, Washington, D.C., 1960), p.8.

53. Conversation with Edmund G. Beacham, Baltimore, Md., January 7, 1981; E.G. Beacham, D.G. Carroll, and F.G. Hubbard, "Baltimore City Hospitals: Toward a Third Century of Progress," *Maryland State Medical Journal* 22 (November 1973):57; *Baltimore Evening Sun,* December 13, 1965; *Baltimore Sun,* December 15, 1965.

54. Conversation with Edmund G. Beacham, January 7, 1981; conversation with Edyth Schoenrich, Baltimore, Md., November 17, 1980.

55. Howard A. Rusk, *A World to Care For* (New York: Random House, 1972), pp. 115–119; Herbert E. Klarman, *Hospital Care in New York City: The Roles of Voluntary and Municipal Hospitals* (New York: Columbia University Press, 1963), p. 44; Helen Eastman Martin, *The History of the Los Angeles County Hospital (1878–1968) and the Los Angeles County-University of Southern California Medical Center (1968–1978)* (Los Angeles: University of Southern California Press, 1979), p. 260; Citizens Research Council of Michigan, *Administrative Survey of the Wayne County General Hospital and Infirmary, Eloise, Michigan* (Detroit, 1954), pp. 42–44.

56. New York City Department of Hospitals, *Research: An Experiment in Municipal Organization, Goldwater Memorial Hospital* (New York, 1945), p. 62; City of Boston, Department of Health and Hospitals, *105th Annual Report* (Boston, 1968), p. 28.

57. S.L. Clemens, *The Adventures of Huckleberry Finn* (New York: Rinehart ed., 1957), p. 80, author's italics.

58. Conversation with Susan Cromer, Washington, D.C., March 1, 1978; Joe Adcock and Cynthia Adcock, "Houston's Hospitals: The Smell of Charity," *Nation* 200 (January 4, 1965):8.

59. Constance McLaughlin Green, *The Secret City: A History of Race Relations in the Nation's Capital* (Princeton, N.J.: Princeton University Press, 1967), pp. 59, 66–67, 111–116; Washington Asylum Hospital, *Report of the Intendant for the Fiscal Year Ended June 30, 1903* (Washington, D.C.: U.S. Government Printing Office, 1903), p. 4.

60. *Washington Post,* November 24, 1947; Dietrich C. Reitzes, *Negroes and Medicine* (Cambridge, Mass.: Harvard University Press, 1958), pp. 190–201; W. Montague Cobb, President, Medico-Chirurgical Society of the District of Columbia, to Guy Mason, Board of Commissioners, District of Columbia, in

*Journal of the National Medical Association* 39 (March 1947):80; *Washington Star,* March 20, 1948, June 3, 1948, August 29, 1956; Conversation with W. Montague Cobb, Baltimore, Md., May 11, 1981.

61. Katherine A. Harvey, "Practicing Medicine at the Baltimore Almshouse, 1828–1850," *Maryland Historical Magazine* 74 (September 1979):225; Helen Merryman Streett, "Hospital and Dispensary Care for the Colored in Baltimore" (M.A. thesis, Johns Hopkins University, 1927), pp. 6, 7; Minutes, Baltimore Board of Supervisors of City Charities, April 7, 1934; Minutes of the Medical Advisory Board, Baltimore City Hospitals, August 8, 1947, January 5, 1950, May 3, 1951, Alan M. Chesney Archives, Johns Hopkins University, Baltimore; Francis Chinard to Harry F. Dowling, May 9, 1980.

62. H.M. Green, "A Brief Study of the Hospital Situation among Negroes," *Journal of the National Medical Association* 22 (July-September 1930):114.

63. "N.M.A. Communications," *Journal of the National Medical Association* 15 (January-March 1923):54; W.H.A. Barrett, "The Missouri Hospital Situation," ibid. 22 (July-September 1930):140–141; Green, "The Hospital Situation among Negroes," pp. 113–114; Green, "Our Hospital Problem," ibid. 27 (May 1935):73–74; J.J. Cary, "The Hospital Unit, Knoxville, Tenn." ibid. 34 (May 1942):131; Editorial, "Our Hospital Problems," ibid. 21 (July-September 1929):114–116; "Professional News," ibid. 43 (September 1951):346; Charles H. Garvin, "The Negro Physicians and the Hospitals of Cleveland," ibid. 22 (July-September 1930):124–126; W. Montague Cobb, "Cleveland's Oak Forest Hospital Celebrates First Birthday," ibid. 51 (March 1959):151. The quotation is from J.E. Levy, "Integration Battlefront," ibid. 50 (January 1958):69.

64. U.S. Department of Commerce, Bureau of the Census, *Statistical Abstract of the United States, 1939* (Washington, D.C.: U.S. Government Printing Office, 1940), pp. 22–23; U.S. Department of Commerce, Bureau of the Census, *County and City Data Book, 1962: A Statistical Abstract Supplement* (Washington, D.C.: U.S. Government Printing Office, 1962), p. 536; Peter Marshall Murray, "The Hospitals in New York State and Harlem," *Journal of the National Medical Association* 22 (July-September 1930):132–133; Arthur T. Davidson, "A History of Harlem Hospital," ibid. 56 (September 1957):373–380.

65. Samuel U. Rodgers, "Kansas City General Hospital No. 2: A Historical Summary," *Journal of the National Medical Association* 54 (September 1962):528, 534; the quotation is on p. 530.

66. James J. McGuire, "Recent Advances at Homer G. Phillips Hospital," *Journal of the National Medical Association* 47 (September 1955):323; H. Phillip Venable, "The History of Homer G. Phillips Hospital," ibid. 53 (November 1961):550; Rodgers, "Kansas City General Hospital No. 2," p. 534; "Kansas City General Hospitals No. 1 and 2 Consolidated," ibid. 50 (March 1958):142.

67. John D. Dingell, "Address," *Journal of the National Medical Association* 54 (May 1952):506; "Texts of the Atlanta and Butler, N.C., Hospital Suits," ibid. 55 (January 1963):51–52; "Atlanta Hospitals," *Urban Health* 7 (September 1978):37–42.

68. Bernard C. Randolph, "The Situation in St. Louis," *Journal of the National Medical Association* 55 (July 1963):343; "Texts of the Atlanta and Butler, N.C., Hospital Suits." The quotation is from a report entitled

"Proceedings: Imhotep National Conference on Hospital Integration," ibid. 49 (November 1957):431.

69. W.M. Cobb, "Integration Battlefront," *Journal of the National Medical Association* 50 (January 1958):72. For other comparisons, see A.M. Townsend, "Integration Battlefront," ibid., pp. 68–69; J.M. Levy, "Integration Battlefront," ibid., p. 69; Rodgers, "Kansas City General Hospital No. 2," pp. 535–539.

70. The Supreme Court upheld a decision of the Fourth Circuit Court requiring two hospitals in Greensboro, N.C., recipients of Hill-Burton funds for construction, to cease denying patients access to the hospital and physicians and dentists positions on the staff on the basis of race. Following this, the Department of Health, Education and Welfare changed its policy to require that "all portions and services of the entire facility" for which Hill-Burton funds were sought "will be made available without discrimination on account of race, creed, or color; and that no professionally qualified person will be discriminated against . . . with respect to . . . professional practice in the facility." Quoted from 42 C.F.R. 53.112 in "The Hill-Burton Program and Civil Rights," *Journal of the National Medical Association* 57 (January 1965):60.

71. James H. Quigley, "Hospitals and the Civil Rights Act of 1964," *Journal of the National Medical Association* 57 (November 1965):445–446; Letter and Guidelines, W.H. Stewart, Surgeon General, Public Health Service, to all hospital administrators, March 4, 1966, in ibid. 58 (May 1966):212–213; Dingell, "Address," p. 505.

72. "Chicago Activity against Hospital Discrimination Continues," *Journal of the National Medical Association* 47 (July 1955):265; Clyde W. Phillips, Jeb Boswell, and Helen Jaffe, "The Inner-City Patient," *Hospitals* 44 (November 16, 1970):55–56.

73. Data on black births in Chicago is from Reitzes, *Negroes and Medicine*, p. 104; and "Negro Births in Chicago, 1954–1955," *Journal of the National Medical Association* 49 (March 1957):124–125. Information on black staff doctors is from Franklin C. McLean, "Negroes and Medicine in Chicago," ibid. 52 (September 1960):369; and Chicago Commission on Human Rights, "Negro Physicians and Medical Students Affiliated with Chicago Hospitals and Medical Schools," ibid. 57 (May 1965):237, 240–242.

74. George C. Sears, "Hospital Administration under the Eighteenth Amendment," *Boston Medical and Surgical Journal* 189 (September 20, 1923):398; Merrill Moore and Mildred Geneva Gray, "The Problem of Alcoholism at the Boston City Hospital," *New England Journal of Medicine* 217 (September 2, 1937):381–388.

75. John J. Byrne, ed., *A History of the Boston City Hospital, 1905–1964* (Boston, 1964), pp.93–94.

76. *Washington Star,* February 12, 1936; *Washington Post,* February 12, 1936; *Report of the Government of D.C., 1947*, p. 203; Council on Medical Education and Hospitals of the American Medical Association, "Hospital Service in the United States," *Journal of the American Medical Association* 137 (August 14, 1947):1403, 1416; "The Social Service Department," *The Record* (Baltimore) 4 (July 1961):4; *Baltimore Sun,* July 1, 1960.

77. Conversation with Burton C. D'Lugoff, Baltimore, Md., February 19, 1980. The quotation is from the remarks of Franklin A. Neva in Boston City Hospital, *The Harvard Medical Unit: An Autobiography, 1915–1964* (Boston, n.d.), p. 46.

78. D. Hayes Agnew, "The Medical History of the Philadelphia Almshouse," in *History of Blockley: A History of the Philadelphia General*

*Hospital from Its Inception, 1731–1928,* comp. John Welsh Croskey (Philadelphia: W.A. Davis, 1929), picture opp. p. 20; Charles E. Rosenberg, "Social Class and Medical Care in Nineteenth-Century America: The Rise and Fall of the Dispensary," *Journal of the History of Medicine and Allied Sciences* 29 (January 1974):32–33, 49–53; John Duffy, *A History of Public Health in New York City, 1625–1866* (New York: Russell Sage Foundation, 1968), p. 510; Michael M. Davis, *Clinics, Hospitals and Health Centers* (New York: Harper and Brothers, 1927), p. 38.

79. American Hospital Association, *American Hospital Directory, 1945* (Chicago, 1945), p. 592; *Hospitals,* Guide Issue 40 (August 1, 1966): 474–475; ibid. 45 (August 1, 1971):488.

80. *Hospitals,* Guide Issue 29 (August 1955):57; ibid. 40 (August 1, 1966):474–475; ibid. 45 (August 1, 1971):488; *Washington Daily News,* October 24, 1967.

81. Some claim that Bellevue's outpatient department opened in 1863, others in 1867. See James J. Walsh, *History of Medicine in New York: Three Centuries of Medical Progress,* 5 vols. (New York: National Americana Society, 1919), 3:725; Henry A. Davidson, "Bellevue: The Country's Oldest Hospital Is a Proving Ground for Some of Medicine's Newest Ideas," *Medical Economics* 26 (March 1949):132. See also Joseph Chapman Doane, "A Brief History of the Philadelphia General Hospital from 1908 to 1928," in Croskey, *History of Blockley,* p. 122.

82. "Baltimore Hospital: Resident-Intern Center," *Resident Physician* 9 (January 1963):78; Alan M. Chesney, *The Hospitals of Baltimore and the Indigent Sick* (Baltimore, 1932), p. 12; Arthur J. Lomas, "The Out-Patient Department," *Transactions of the Medical and Chirurgical Faculty of the State of Maryland, 1937,* p. 115; Beacham et al., "Toward a Third Century of Progress," p. 46; Carroll, "History of Baltimore City Hospitals," September 1966, p. 106; McMillin, "Baltimore City Hospitals," p. 752.

83. *Washington Post,* October 19, 1945; "News and Personals," *Medical Annals of the District of Columbia* 16 (November 1947):640.

84. MacEachern, "Summary Report of Findings," pp. 37, 138; and Board of Commissioners of Cook County, "Summary of Medical Activities in Cook County Hospital," *Chicago Hospital Council Bulletin* 1 (October 1938):14. See also "Report of Committee on Survey of Cook County Hospital," p. 15; Karl A. Meyer, "Historical Background of Cook County Hospital," *Quarterly Bulletin of Northwestern University Medical School* 23 (Fall 1949):273–274; "Medical News," *Journal of the American Medical Association* 177 (August 19, 1961):30.

85. Boston City Hospital Board of Trustees, *First Medical and Surgical Report of the Boston City Hospital,* ed. J. Nelson Borland and David W. Cheever (Boston: Alfred Mudge and Son, 1870), pp. 21, 22; Byrne, *History of Boston City Hospital,* p. 26; Morris J. Vogel, *The Invention of the Modern Hospital: Boston, 1870–1930* (Chicago: University of Chicago Press, 1980), p. 88; David W. Cheever et al., *A History of the Boston City Hospital from Its Foundation until 1904* (Boston: Municipal Printing Office, 1906), pp. 125, 159.

86. Board of Trustees and Staff of the Boston City Hospital, *Report of the Joint Committee,* p. 39; "The Dedication of the New Outpatient Department of the Boston City Hospital," *Boston Medical and Surgical Journal* 191 (October 30, 1924):845–847; "Medical News: Outpatient Department— Boston City Hospital," *Journal of the American Medical Association* 89 (August 13, 1927):530; Maxwell Finland to Harry F. Dowling, and accompanying

material, January 22, 1981, April 7, 1981; "Fifth and Sixth (BU) Medical Services," in City of Boston, Hospital Department, Annual Report, 1963 (mimeographed), n.p.

87. Dickinson W. Richards, "The Hospital and the City," in *Training of the Physician: Papers Presented as Part of the Celebration of the Centennial of the Boston City Hospital and the Fortieth Anniversary of the Thorndike Memorial Laboratory* (Boston, 1964), p.127.

88. See, for instance, De Hartog, *The Hospital*, pp. 36–40.

89. "Open Outpatient Clinic at Gallinger Hospital," *Modern Hospital* 70 (January 1948):170; *Washington Times-Herald*, January 7, 1951; *Washington Post*, May 10, 1961.

90. *Washington News*, February 14, 1956.

91. *Washington Post*, May 10, 1961, September 14, 1968; Government of the District of Columbia, *Annual Report, 1966* (Washington, D.C.: U.S. Government Printing Office, 1966), pp. 4–7; District of Columbia, Department of Public Health, *Vistas to Public Health in the District of Columbia: A Report to the People* (Washington, D.C., 1965–66).

92. Chicago Board of Health, *Preliminary Report on Patterns of Medical and Health Care in Poverty Areas of Chicago and Proposed Health Programs for the Medically Indigent* (Chicago, 1966), p. 33; John D. Porterfield, Director, Joint Commission on Accreditation of Hospitals, to W.M.McCoy, Administrator, Cook County Hospital, accompaniment to letter of April 19, 1968, p. 4.

93. Robert H. Hamlin, Arnold I. Kisch, and H. Jack Geiger, "Report on the Boston City Hospital and Boston Health Department to Mayor John F. Collins and the Board of Trustees of Boston City Hospital" (mimeographed, Boston, 1965), available at Francis A. Countway Library of Medicine, Boston; James B. Sacchetti, "108-Year-Old Boston City Hospital Undergoing Extensive Modernization," *Hospital Topics* 50 (June 1972):52, 72.

94. Marion M. Wolk, "Know Your Hospital Day: A Vital Quiz for *All* Marylanders" (mimeographed, Baltimore, 1967), copy in Mayor's Administrative Files, Record Group 9, Baltimore City Archives.

95. An example of how the emergency service taxed the hospital's facilities was shown in a study of the effectiveness of patient care in its emergency room that was carried out by interviewing patients several months after their initial visit. It was found that patients were not always examined completely; only 68 percent of X-ray examinations were completed (in part because some examinations could not be scheduled for two to three months); only 55 percent of patients had an overall evaluation; and only 37 percent of patients who were found to have abnormalities on X-ray examination were receiving appropriate therapy. The evaluators concluded that only 27 percent of the sample of patients studied had received effective medical care. Robert H. Brook and Robert L. Stevenson, Jr., "Effectiveness of Patient Care in an Emergency Room," *New England Journal of Medicine* 283 (October 22, 1970):904–907. These unimpressive results were not unique to the Baltimore City Hospitals; care was probably much better there than in many other public or private hospitals. They merely reflected the unpreparedness of hospitals in general and city hospitals in particular for the invasion of their emergency rooms by a host of patients.

96. Citizens Board of Inquiry into Health Services for Americans, *Heal Yourself*, 2d ed. rev. (Washington, D.C.: American Public Health Association, 1972), pp. 106–111.

97. "Interim Report on Medical Care for Baltimore City," pp. 93–114; "State Planning Commission Reports on Medical Care Program," *Baltimore*

*Health News* 30 (March-April 1953):98–110; Harry L. Chant, "How the Hospitals Cooperate: Baltimore's Experiment in Indigent Medical Care," *Hospitals* 22 (October 1948):55; Alexander, "Evolution of Public Medical Care in Baltimore," p. 65; television interview with health commissioner, December 21, 1961, in *Baltimore Health News* 39 (February 1962):13–14.

98. American Medical Association Study of Medical Care, "Report of Committee on Medical Economics of the Chicago Medical Society," *Journal of the American Medical Association* 113 (July 8, 1939):151; U.S. Public Health Service, *The Chicago-Cook County Health Survey* (New York: Columbia University Press, 1949), pp. 1089, 1108; "Welfare Council Group Recommends Branch for Cook County Hospital," *Modern Hospital* 84 (May 1955):178; Edward C. Banfield, *Political Influence* (New York: Free Press, 1961), pp. 15–17, 25–29.

99. Robb K. Burlage, *New York City's Municipal Hospitals: A Policy Review* (Washington, D.C.: Institute for Policy Studies, 1967), pp. 121–124.

100. Editorial, *Washington Star,* September 17, 1939; conversation with Sol Katz, Baltimore, Md., June 1, 1981.

101. Remarks of the Associate Director for Hospitals of the District of Columbia, in U.S. Senate, Committee on the District of Columbia, Subcommittee on Public Health, Education, and Welfare, *District of Columbia Hospital Construction and Modernization,* p. 58.

## 8. Problems, Protests, and Possible Solutions

Epigraphs: Ray E. Brown, "The Plight of the Public Hospital: 1. The Public Hospital," *Hospitals* 44 (July 1, 1970):40; "The Public Hospital—End of the Line?" *American Medical News,* November 30, 1970, p. 8.

1. See, for instance, *Washington Star,* September 9, 1969; *Washington Post,* September 10, 1969, April 29, 1972, May 7, 1972; "Measures Approved to Aid Crowded County Hospital," *Hospitals* 44 (February 16, 1970):44. The quotation is from William Shakespeare, *King Lear,* act 1, sc. 2, line 125.

2. Michael G. Michaelson, "Medical Students: Healers Become Activists," *Saturday Review,* August 16, 1969, pp. 41–43, 53.

3. A. Tetelman, "Public Hospitals—Critical or Recovering?" *Health Service Reports* 88 (April 1973):296.

4. "Doctors Demand Cook County Hospital Budget Increase; Board Proposes $2.1 Million Hike," *Modern Hospital* 106 (February 1966):186.

5. The newspaper advertisement is reported in "House Staff's Public Protest Aids Bill to Change Cook County Hospital," *Modern Hospital* 112 (April 1969):36. See also Report of the Illinois Legislative Investigating Commission, *Cook County Hospital* (Chicago, 1972), p. xii; "Cook County Hospital Director Resigns to Bring 'Real Issue' into Focus," *Modern Hospital* 114 (June 1970):36c–36d; "1,700 Patients without a Hospital," *Medical World News* 11 (June 5, 1970):15–16; Robert J. Freeark, "The Plight of the Public Hospital. Case Study: Chicago," *Hospitals* 44 (July 1, 1970):57–60.

6. "Cook County Starts New Life under New Commission," *Modern Hospital* 115 (August 1970):39; "Cook County Post Goes to Dr. Haughton, New York Official," ibid. 115 (October 1970):42; "One-Year Accreditations Given Two Major Municipal Hospitals," *Hospitals* 44 (May 16, 1970):119–120; Joel H. Goldberg, "The Cook County Lesson: How to Cripple a Great Hospital," *Hospital Physician* 8 (May 1972):27, 31; Joel H. Goldberg, "The Power Struggle at Cook County Hospital Is Boiling Over," ibid. 8 (February 1972): 31–32; Rolf M. Gunnar to Harry F. Dowling, May

29, 1981; Joel H. Goldberg, "The Ego Trip Is Over at Cook County," *Hospital Physician* 9 (January 1973):57–62; Robert M. Cunningham, Jr., "Strike," *Modern Healthcare* 5 (February 1976):41–48; Donald F. Phillips, "Administrative Prerogatives Challenged at Hearings," *Hospitals* 50 (April 16, 1976):57–59; "Dr. Young's Rehiring Upheld," *American Medical News*, August 29, 1977, pp. 1, 13; "Board Drops Bid to Fire Dr. Young," ibid., October 3, 1977, p. 14.

7. Judy Alsofrom, "Cook County Hospital Nears Bankruptcy," *American Medical News*, November 16, 1979, pp. 3, 11; "Cook County Hires Hospital Managing Firm," ibid., January 25, 1980, p. 9.

8. Editorial, "A Century of Service," *New England Journal of Medicine* 270 (May 28, 1964):1198–1199.

9. The quotation is cited in "Boston City Hospital on the Critical List," *Medical World News* 9 (June 7, 1968):9. See also *Boston Morning Globe*, May 16, 1968; Boston Department of Health and Hospitals, *Progress Notes*, Winter 1970, n.p.; *Boston Phoenix*, December 1, 1971.

10. Editorial, "Report on Boston's Health Services,"*New England Journal of Medicine* 273 (July 1, 1965):49–50; Editorial, "Boston's Department of Health and Hospitals," ibid. 274 (March 24, 1966):687–689; Robert H. Hamlin, Arnold I. Kisch, and H. Jack Geiger, "Report on the Boston City Hospital and Boston Health Department to Mayor John F. Collins and the Board of Trustees of Boston City Hospital" (mimeographed, Boston, 1965); Robert H. Hamlin, Arnold I. Kisch, and H. Jack Geiger, "Administrative Reorganization of Municipal Health Services—The Boston Experience," *New England Journal of Medicine* 273 (July 1, 1965):26–29; Boston Department of Health and Hospitals, *105th Annual Report* (Boston, 1968), pp. 8–9; *Boston Herald Traveler*, March 31, 1970.

11. *Boston Phoenix*, December 1, 1971, December 8, 1971; *Boston Morning Globe*, December 28, 1972, March 1, 1973; *Boston Herald Traveler and Record American*, January 4, 1973; Boston Department of Health and Hospitals, *Progress Notes*, December 1975, p. 7; *Harvard Crimson*, February 23, 1973; *Boston Evening Globe*, February 28, 1973.

The last class graduated from the nursing school in 1975. In 1974 a baccalaureate program in nursing began at the Boston City Hospital under the auspices of Boston State College. Conversation with Anne G. Hargreaves, Boston, Mass., December 4, 1980.

12. *Washington Post*, August 18, 1977, February 19, 1978.

13. Ibid., June 19, 1980.

14. Illinois Legislative Investigating Commission, *Cook County Hospital*, p. 16; Helen Eastman Martin, *The History of the Los Angeles County Hospital (1878–1968) and the Los Angeles County-University of Southern California Medical Center 1968–1978)* (Los Angeles: University of Southern California Press, 1979), pp. 517–520.

15. "Cornell University Severs Its Long-Standing Association with New York's Bellevue Hospital," *Modern Hospital* 107 (December 1966):176; Charles A. Ragan, Jr., "The End of the Columbia Medical Division at Bellevue," *Resident Physician* 13 (April 1967):59–62; "$27,500,000 Medical Center Planned by New York U. and City," *Hospital Management* 61 (February 1946):26–28; Illinois Legislative Investigating Commission, *Cook County Hospital*, pp. 113–114.

16. Boston Department of Health and Hospitals, *Report by the Board of Health and Hospitals to Mayor Kevin H. White* (Boston, 1968), p. 12; Maxwell Finland to Harry F. Dowling, and accompanying material, May 7, 1981;

*Pawtucket* [R.I.] *Times,* March 6, 1973; *Boston Herald American,* March 18, 1975; conversation with John J. Byrne, Boston, Mass., December 4, 1980; conversation with Anne G. Hargreaves, Boston, Mass., December 4, 1980; conversation with William McCabe, Boston, Mass., December 4, 1980; Ephraim Friedman, "Annual Discourse—The Boston City Hospital: A Tale of Three 'Cities,'" *New England Journal of Medicine* 289 (September 6, 1973):503–506.

17. Dean George P. Berry, Harvard Medical School, "Remarks at Boston City Hospital, December 10, 1964," in *Training of the Physician: Papers Presented As Part of the Celebration of the Centennial of the Boston City Hospital and the Fortieth Anniversary of the Thorndike Memorial Laboratory* (Boston, 1964), p. 118.

18. *Washington Post,* April 16, 1974.

19. "Harborview Medical Center, 1877–1977: A Century of Community Service and Caring" (pamphlet, Seattle, 1978), p. 10; Martin, *History of Los Angeles County Hospital,* pp. 224–226; R.E. Tranquada and R.F. Maronde, "The Hospital within a Hospital: An Empirical Experiment in Health Care in a Major Metropolitan Hospital," *Bulletin of the New York Academy of Medicine* 48 (April 1972):560; Robb K. Burlage, *New York City's Municipal Hospitals: A Policy Review* (Washington, D.C.: Institute for Policy Studies, 1967); Howard D. Young and Alex Rosen, Design and Evaluation, Inc., *New York City's Municipal Hospital System: Physicians' Perceptions,* Study Commissioned by Society of Urban Physicians (New York, 1972), p. 21; Ray E. Trussell, "Current Efforts in the Municipal Hospitals," *Bulletin of the New York Academy of Medicine* 43 (March 1967):211–218.

20. "Independent Body To Run New York City Hospitals," *Hospitals* 43 (June 16, 1969):43; "The Public Hospital—End of the Line?" p. 9; "New York's Crumbling Public Hospital 'Empire' Gets Confident New Czar," *American Medical News,* April 28, 1978, p. 13.

21. Alexander Pope, "An Essay on Man," Epistle III, lines 303–304.

22. Morris J. Vogel, "Machine Politics and Medical Care: The City Hospital at the Turn of the Century," in *The Therapeutic Revolution: Essays in the Social History of American Medicine,* ed. Morris J. Vogel and Charles E. Rosenberg (Philadelphia: University of Pennsylvania Press, 1979), p. 173; J.C. Lashof, "The Health Care Team in the Mile Square Area, Chicago," *Bulletin of the New York Academy of Medicine* 44 (November 1968):1363–1369; J.E. Fine, "The Organization of Primary Health Care Services: The Boston Experience," in *Public General Hospitals in Transition— 1. Partnership in Planning: The Boston Scene.* American Rehabilitation Foundation, Institute for Interdisciplinary Studies, Health Services Research Center (Minneapolis, 1970), pp. 13–19.

23. Citizens Board of Inquiry into Health Services for Americans, *Heal Yourself,* 2d ed. rev. (Washington, D.C.: American Public Health Association, 1972), pp. 110–116; Illinois Legislative Investigating Commission, *Cook County Hospital,* p. 121.

24. Harmon T. Rhoads, Jr., and Joseph B. Cleary, "New York Medical College and Metropolitan Hospital Center," *New York State Journal of Medicine* 76 (April 1976):596; Herbert E. Klarman, *Hospital Care in New York City: The Roles of Voluntary and Municipal Hospitals* (New York: Columbia University Press, 1963), p. 34; *Washington Post,* January 8, 1979.

25. David L. Corwin, "The Plight of the Public Hospital. Case Study: Denver," *Hospitals* 44 (July 1, 1970):61–64.

26. Tranquada and Maronde, "Hospital within a Hospital," pp. 565–576.

27. *Baltimore Evening Sun,* December 26, 1978, December 27, 1978, December 29, 1978, December 30, 1978; Minutes of the 87th Meeting of the Hospital Commission, Baltimore City Hospitals, July 12, 1973, Department of Legislative Reference Files, Record Group 29, Baltimore City Archives.

28. Monroe Lerner and Zili Amsel, "A Study of the Role of Baltimore City Hospitals in the Community" (mimeographed, Baltimore, 1970), pp. i–iv.

29. Burlage, *New York City's Municipal Hospitals,* pp. 427–430; Minutes of the 88th Meeting of the Hospital Commission, Baltimore City Hospitals, August 16, 1973, Department of Legislative Reference Files, Record Group 29, Baltimore City Archives; Conversation with Edmund G. Beacham, Baltimore, Md., November 6, 1980.

30. Report of the Commission on Public-General Hospitals, *The Future of the Public-General Hospital: An Agenda for Transition* (Chicago: Hospital Research and Educational Trust, 1978), p. 22.

31. *Baltimore Sun,* May 25, 1980, November 5, 1980; conversation with Edmund G. Beacham, November 6, 1980.

32. John Craig and Michael Koleda, "The Urban Fiscal Crisis in the United States, National Health Insurance, and Municipal Hospitals," *International Journal of Health Services* 8, no. 2 (1978):331; "Boston City Hospital Crisis: 'Nightmare' Finance Dilemma," *Medical Tribune,* December 2, 1968, pp. 1, 23; Martin Cherkasky, "A Hospital Administrator Says that the City Should Get Out of the Hospital Business," *New York Times Magazine,* October 8, 1967.

33. Craig and Koleda, "The Urban Fiscal Crisis," p. 338.

34. The quotation is from "N.Y. Municipal Hospitals Optimistic of Future As Civic Group Asks End of 'Charity Medicine,'" *Hospitals* 42 (July 1, 1968):110. See also Ralph A. Milliken and Dennis D. Pointer, "Organization and Delivery of Public Medical Care Services in New York City," *New York State Journal of Medicine* 74 (January 1974–Part 1):108; Jeff Ryser, "Health Fund Cuts Pushing New York to Brink of Crisis," *American Medical News,* June 14, 1976, pp. 1, 15; "New York City's Troubled Hospital System," *Urban Health* 6 (February 1977):27.

35. David Riesman, "How the New Blockley Came into Being," *Medical Life* 40 (March 1933):143; "The Closing of Philadelphia Hospital," *Urban Health* 7 (November 1978):40–47; Emily Friedman, "The End of the Line: When a Hospital Closes," *Hospitals* 52 (December 1, 1978):69–75. See also "Philadelphia General Hospital from the Viewpoint of the City Administration," *Philadelphia Medicine* 72 (July 1976):289–291; "Unions Bid to Avert Public Hospital Closing," *American Medical News* 19 (May 17, 1976):19.

36. Sylvester E. Berki, *Hospital Economics* (Lexington, Mass.: Lexington Books, D.C. Heath and Company, 1972), p. 208; "House Staff Protesting Lack of Funds and Support for D.C. General," *Modern Hospital* 113 (November 1969):27–28.

# Index

Adamson, Sarah, 61

Addams, Jane, 38, 54

Administration and administrators of hospitals: protests against, 16–17, 76, 174–176; in almshouse hospitals, 19; professionalization of, 40–41; at Boston City Hospital, 40–41, 53, 104; at Cook County Hospital, 41, 53, 54; at Philadelphia General Hospital, 41–42; at Bay View Asylum and Baltimore City Hospitals, 42, 101, 103; at Washington Asylum and Gallinger Municipal Hospital, 42; in relation to house officers, 53–54; in twentieth century, 104–106. *See also* Governance of city hospitals

Affiliations with medical schools, *see* Medical schools

Alcoholics and alcoholism, *see* Patients in city hospitals

Almonester y Roxas, Don Andres, 34

Almshouses and almshouse hospitals: in England, 7–8; evolving into hospitals, 10, 11, 13, 35; care of patients and nursing in, 14–15, 17–18, 19; attending and resident doctors in, 14–17; graft in, 18–19; medical students in, 19–21; in Midwest and Far West, 35; outpatient services in, 163. *See also entries for individual hospitals and almshouses*

American College of Surgeons: and accreditation of hospitals, 94, 95, 96, 121, 147; program for evaluating hospitals, 107–108, 117, 215n64

American Medical Association and approval of hospitals, 96, 113

Amphitheater clinics, 63–65

Ancker Hospital (St. Paul, Minn.), 139

Antisepsis and asepsis, 27, 128

Apprentices, 19–20

Association of American Physicians, 46, 55

Baltimore: population of, 25–26, 87–89; Department of Hospitals, 104; Department of Public Welfare, 105, 126, 147, 151, 152; Board of Supervisors of City Charities, 114; Commission of Governmental

Efficiency and Economy, 147; private hospitals in, 197n28

Baltimore City Hospitals (including Baltimore Almshouse, Bay View Asylum): beginnings of, 11, 13, 32–33, 37; administration of, 13, 32–33, 42, 101, 103, 104; medical staff of, 15, 113–117, 125–126, 152, 183–184; surgical operations in almshouse, 17; and medical education, medical schools, and medical students, 20–21, 113–117, 125–126, 135, 180, 183; physical plant and facilities of, 32–33, 101–102, 103–104, 146; and chronically ill patients, 33, 153–155, 163; black patients in, 33, 156, 158; governance of, 37, 42, 185; pathology service, 59, 60; costs of patient care and expenditures, 77, 91, 102–103, 104, 149–150; admissions, births, and patient census, 91, 104; bed capacity and crowding, 91, 101, 104, 148, 177; and politics and politicians, 100, 101, 102–103; nursing and nursing school of, 101, 140–141, 144, 155; X-rays and electrocardiograms in, 102; accreditation of, 108; Medical Advisory Board, 116, 158; research in, 131; house officers in, 133, 135; paying patients in, 151–152; social concerns and social services, 162, 163; outpatient services in, 165, 168, 232n95; community services of, 169, 183–184; Chesapeake Physicians Professional Association, 183–184

Bay View Asylum, *see* Baltimore City Hospitals

Bellevue Establishment, *see* Bellevue Hospital

Bellevue Hospital (including Bellevue Establishment, New York Almshouse): evolution of, 11, 12, 27; bed capacity and crowding, 18, 32, 78; medical students in, 20, 135; physical plant and facilities of, 32, 138, 146; governance of, 36, 37; medical staff of, 49, 110–111, 122; house officers, 52; specialty services, 56–57, 59; care of pa-

# Index

Chicago, University of, medical school, 53, 72–73. *See also* Cook County Hospital

Chicago Commission on Human Relations, 161

Chicago Hospital for Women and Children, 80

Chicago Medical School, 122

Chicago Medical Society, 120

Children's Hospital (Boston), 80

Children's Hospitals (New York City), 79

Chinard, Francis, 117

Christian, Henry, 66

Chronically ill patients, 33, 153–156, 163

Churchill, Edward D., 112

Cincinnati General Hospital, 35, 139

Cities in the United States: beginnings of, 9; growth of, 25–26, 87, 88; governments of, 35, 44; kinds of people in, 87, 89. *See also entries for individual cities*

City Hospital (Jersey City), 139

City Hospital (New York City), 79

Civil Rights Act of 1964, 160

Civil Service examinations, 47, 52–53, 121

Cleveland Metropolitan Hospital (including Cleveland City Hospital), 3, 135, 159

Clinical clerkships, 66–67, 135. *See also* Medical education

Coleman, George H., 132

College of Medicine of Maryland, 20

College of Physicians and Surgeons (Baltimore), 114

College of Physicians and Surgeons (New York City), 20, 59, 64

Collins, Patrick, 39

Colorado, University of, medical school, 183

Columbia Hospital for Women (Washington, D.C.), 33, 74

Columbia Point Clinic (South Boston), 169

Columbia University medical school, 131, 178. *See also* Bellevue Hospital; College of Physicians and Surgeons (New York City)

Commission on Public-General Hospitals, 184

Community relations of city hospitals, 162, 163, 167, 181–184

Connecticut State Hospital, 71

Contagious diseases, *see* Infections and infectious diseases

Cook County Hospital: beginnings of, 27–29; physical plant and facilities of, 28–29, 78, 79–80, 91, 93–97, 145–146, 166, 174; governance of, 36–39, 95, 174–175; administration of, 41, 95–99, 104, 123; medical staff of, 46–49, 50, 120–123, 219n35; house officers in, 52–53, 174–175; specialty services in, 56, 57; homeopaths and eclectics in, 61; women doctors in, 62; medical schools and, 62, 120–123; medical education in, 63, 64, 65, 135; nursing and nursing schools in, 71–73, 94, 138–140, 143–145; expenditures by, 77–79, 91, 93, 95, 102, 140; bed capacity and crowding, 78–79, 89, 91, 93, 94, 148, 177; and the community, 78–79, 96, 97, 137, 169; social services in, 80–81; attitudes toward patients, 80–81, 138; admissions, census, and births, 91, 93, 94; accreditation of, 94, 95, 96, 108, 121, 147; subprofessional employees, 95; investigations, criticisms, and protests, 96–97, 147, 174–175; research in, 130–131, 132; blacks in, 161; outpatient services in, 165–166; Health and Hospitals Governing Commission, 174–175

Cook County School of Nursing, 73, 143

Cornell University medical school, 178. *See also* Bellevue Hospital

Costs of care in hospitals, 77, 91, 102, 149, 150. *See also entries for individual hospitals*

Councilman, William T., 59–60, 114

Cournand, André, 132

Cowles, Edward, 41

Curley, James Michael, 39, 90–93, 106, 130, 135, 146

Cushing, Richard Cardinal, 105

Cutler, Rev. Manasseh, 14, 17

Da Costa, Jacob M., 64

Daley, Richard, 106–107

Davidson, Charles, 167

Denver General Hospital, 35, 183, 186

Detroit Receiving Hospital, 139

DeVise, Pierre, 95

Dickens, Charles, 8, 54

Disaster syndrome, 148, 174

Dispensaries, 163–164. *See also* Outpatients and outpatient clinics

District of Columbia, *see* Washington, D.C.

# Index

Harlem Hospital (New York City), 79, 132, 149, 159, 178, 214n45

Harvard University medical school: pathology department of, 60; and Boston City Hospital, 62, 63, 66, 109–113; and clinical teaching, 66. *See also* Boston City Hospital

Haughton, James G., 174

Health and Hospitals Governing Commission (Cook County, Illinois), 174–175, 181

Hektoen, Ludwig, 53

Hektoen Institute for Medical Research, 130–131

Henricopolis almshouse, 10

Herrick, James B., 48, 52, 64

Hill-Burton Act, 160

Hispanic Americans, 161

Hobson, Elizabeth C., 71

Homeopathy, 60–61

Homer G. Phillips Hospital (St. Louis), 159–160

Hosack, David, 15

Hospitals: definitions of "public-general," "city," "private," "voluntary," and "proprietary," 1–2; in England, 7–9; in North American colonies, 9; evolving from almshouses, 10–13; infections in, 26–27, 30–31, 69; architecture of, 27; growth of, 27, 78–79, 91, 101, 148–149; in Chicago, 28–29, 195n9; in Boston, 29–30; in New York City, 32, 36, 126–128, 169, 178; in Washington, D.C., 33; special, for infectious diseases, 57; relationships of, with medical schools, 62–63, 123–125; subprofessional employees of, 92–93; accreditation of, 107–108; in Baltimore, 197n28. *See also* Administration and administrators of hospitals; Costs of care in hospitals; Governance of city hospitals; *entries for individual hospitals*

Hospitals, private, 27, 77, 149. *See also entries for individual hospitals*

Hospitals, public, 27, 77, 149. *See also entries for individual hospitals*

Hospitals, voluntary, 77–78, 187–188

Hospital service, definition of a, 197n26

Hospital Survey and Construction Act, 160

House officers: in almshouse hospitals, 16–17, 20, 21; protests against hospital administration by, 16–17, 76, 133, 174–176; in city hospitals, 51–54, 132–135; at Cook County Hospital, 52–53, 121–122; at Bellevue Hospital, 52; at Gallinger Municipal Hospital, 118; in New York municipal hospitals, 127; length of internships of, 132–133; attitudes of, toward patients, 133–134, 157, 163; exploitation of, 134–135

Howard University medical school, 157–158. *See also* District of Columbia General Hospital

Hussey, Hugh, 119, 136

Illinois, University of, medical school, 29, 122. *See also* Cook County Hospital

Illinois Training School for Nurses, 72–73

Immigrants, 26, 38, 87–89, 105, 156

Infant's Hospital (New York City), 79

Infections and infectious diseases: spread of, 26–27; hospitals for, 28, 38, 58–59, 100; special services for, 57–59

Insurance, medical, 103

Interns, *see* House officers

Irish: in Boston, 29, 90, 105; in politics, 39–40; on hospital staffs, 46; alcoholism among, 161–162

Janeway, Edward G., 49, 59

Janeway, Theodore C., 49

Jefferson Davis Hospital (Houston), 87, 133–134

Jefferson Medical College, 49, 51, 64. *See also* Philadelphia General Hospital

Jeghers, Harold, 162

John McCormick Institute for Infectious Diseases, 130

Johns Hopkins Hospital, 66, 77, 102, 103, 165, 169, 183

Johns Hopkins Medical School, 49, 50, 101, 111, 114–117, 125–126, 163, 183. *See also* Baltimore City Hospitals

Johns Hopkins University, 105

Johnson, Andrew, 157

Joint Commission on Accreditation of Hospitals, 147. *See also* American College of Surgeons

Kansas City (Missouri) Hospital No. 2, 159, 160

King County Hospital (Seattle), *see* Harborview Hospital

# Index

# Index